Updated and Revised Edition

Generative Ontology and Closure Dynamics (GOCD)

A Background-Free Event Theory

Kaisheng Li and Longji Li

2026

This work is a **pre-publication academic manuscript**. The authors reserve the right to revise, expand, or withdraw any part of the text in future editions.

The views expressed in this book are solely those of the authors and do not necessarily reflect the views of any institution or organization.

June 6, 2026

Acknowledgements

This book represents the outcome of a long intellectual pursuit and is dedicated to my parents, in particular to my late father, to whom it is offered as a belated answer to questions that shaped my life.

More than six decades ago, my parents gave me life. Over the course of those years, life itself gradually became my most persistent problem. In seeking to understand why life arises and why it is finite, I was led—often unexpectedly—from questions of biological existence into the foundations of physics, and eventually toward the deepest structural assumptions underlying contemporary physical theory. This path, which was neither planned nor efficient, occupied much of my adult life. That it could be followed at all is owed not to professional obligation, but to a sustained and uncompromising fascination with physics itself.

I wish to acknowledge my late father. Although he is no longer present, the curiosity and intellectual stubbornness he instilled in me remain decisive influences. This work is written in the hope that, in whatever sense remains meaningful, it may reach him.

I am deeply grateful to my mother for her patience, care, and unwavering support. Her constancy has provided a form of stability without which extended intellectual work of this kind would not have been possible.

My special thanks go to my wife, Yan Zhao. Throughout a project whose motivations and persistence are not always easy to explain, she has been an unfailing companion. By sustaining the practical and emotional conditions of everyday life with remarkable generosity, she made possible the extended periods of concentration on which this book depends. Without her, this manuscript would not exist in its present form.

Finally, this book is offered to readers who retain a serious and undiminished curiosity about the nature of life and the structure of the universe. Whether or not definitive answers can be reached, the act of sustained questioning remains, in itself, a meaningful response to the limits of a finite existence.

Kaisheng Li
May 2026

Contents

III. Closure Dynamics: How Existence Appears 141

A Guide for the Skeptical Physicist

0.1. You Have Every Reason to Remain Skeptical

If you are a rigorously trained physicist, you will probably approach this book with reservations even before opening it. Such reservations are rational, well founded, and entirely justified: this book does not begin with any experimentally validated dynamical equation; it does not take a differentiable manifold as its fundamental stage; for substantial portions of the discussion, it deliberately avoids the familiar topics of "experimental fitting," "precise prediction," and "modifications to the Standard Model" on which you may rely when judging whether a work belongs to physics; and in many places, it may appear to discuss ontology rather than perform calculations.

If this leads you to question whether the book still counts as physics, your skepticism is not prejudice but trained judgment. This book does not seek to eliminate that skepticism. Its purpose is to ask you to direct it toward the right question—not why the book does not resemble physics, but where it can fail. The former is a dispute over style; only the latter is a dispute about physics. The purpose of this preface is to help you make that transition.

0.2. This Book Does Not Require You to Accept Any Conclusion

First, set aside one burden: you do not need to accept any substantive claim made in this book. You may reject the view that the world is constituted by events rather than objects, that probability is a product of readout truncation, that continuity should be suspended, or that spacetime is merely an interface description formed after generation. This book does not require you to stand on its side when you have finished reading it.

It asks only that, at the outset, you temporarily accept a weaker yet stricter premise: this book is not modifying existing physics but auditing its starting points. Its concern is not which equation is more accurate, but which premises have never been permitted to fail. This distinction is the pivot of the entire book. A theory may be extraordinarily precise within its container while remaining silent about where that container came from; the object of the audit is not its precision but its silence—the things silently treated as starting points and thereby exempted from examination. If you can accept that the starting points themselves should also be tested, you already occupy the position this book asks you to take, even if you ultimately reject every answer it offers.

0.3. If You Expect "New Equations," You Will Be Disappointed—At Least in the First Half

This book does not initially present field equations that can be used directly for numerical calculation, nor does it provide an immediately executable list of experimental predictions. This is not because the book disregards experiment, but because it adopts a definite judgment: when the ontological starting point of a framework has not yet been clarified, more precise equations merely conceal an incorrect starting point more securely. Pursuing computational precision before auditing the starting point is not rigor; it only pushes the problem deeper and hides it more effectively.

Accordingly, if you are accustomed to evaluating a physics book by asking whether it can calculate a new cross section, fit an existing curve, or provide a numerical correction for a particular experiment, the first half of this book will not satisfy you. But if you are willing, temporarily, to replace those questions with another set—why must these calculations be performed against a continuous background? Why may probability serve as a starting point without a generative mechanism? Why has "existence" itself never been adjudicated but merely assumed?—then you have entered the level at which this book actually operates.

This does not mean that the book will never derive an equation. In fact, it will: in Chapter 22 and Appendix I, the book shows that, when the discrete mechanisms of the generation layer are coarse-grained under finite-resolution readout, they are compatible with a formally complete continuous effective equation. That equation, however, explicitly belongs to a downstream effective description at the readout layer, rather than to the ontological dynamics of the generation layer—the book insists on introducing no continuous equations into the generation layer because continuity is precisely what it seeks to explain. In other words, the book will ultimately give you an equation, but it will first tell you what that equation is, what it is not, and from where it emerges.

0.4. This Book Is Not "Against Mathematics"; It Opposes Mathematical Premises That Cannot Fail

A crucial clarification is required here; otherwise, the entire book may easily be misread as adopting an anti-formalist position. This book does not oppose mathematical rigor, nor does it oppose the use of limits, continuity, symmetry, or probability as tools. What it opposes is placing these structures in positions where they can never fail.

In modern theories, many key structures effectively exist under "exemption clauses": a continuous limit always seems to exist; spacetime always seems already to be given; probability always seems available for invocation without explanation; and singularities are often deferred to the "theoretical boundary." These are not tested conclusions but prior exemptions—they are assumed by default and are therefore never required to answer the question, "If you are wrong, how would the world refute you?" The objective of this book is not to overthrow these tools, but to revoke their exemptions: every structure treated as fundamental must specify its own failure conditions. A tool genuinely enters the testing system of physics only when it can state the conditions under which it does not hold; otherwise, however elegant it may be, it remains an unaudited assumption.

This requirement is condensed into a set of explicit failure criteria, formally presented in Part Zero and applied throughout the book. Here, only the most direct and operational criterion is offered as an example, so that you may know in advance that the book takes its commitments seriously: if probability is truly a product of readout truncation, then, while the generative configuration remains unchanged, systematically increasing the phase fidelity of readout, shortening the gating interval, and widening the phase-sensitive bandwidth must produce a calculable change in the statistical distribution. If, throughout an experimentally realizable range, the distribution remains completely insensitive and strictly invariant under these changes in the readout protocol, then the book's central claim concerning probability is directly falsified. This is an experimentally operational criterion with order-of-magnitude thresholds—whose measurable proxies include coherence time, visibility, and the CHSH parameter—not a belief capable of accommodating every possible outcome.

0.5. How You Should Read This Book

Do not treat it as an alternative theory. It is neither a competitor to the Standard Model nor a candidate theory of quantum gravity. It is a pre-theoretical framework of generation that attempts to answer several questions arising before any equation appears: what may be called physical existence? What kinds of processes qualify for inclusion in dynamics? What kinds of structures must bear the responsibility of failure? Its relationship to existing physics is complementary rather than substitutive—it does not seek unification at the level of equations, fields, or dynamical variables, but attempts to reassign the ontological status of the entities and structures presupposed by existing theories. In the continuous and large-number limits, existing effective theories should reappear as downstream approximations of this framework; the book does not abolish them, but asks where they come from.

Focus your attention on the "failure conditions." While reading, do not first search for where the book "explains phenomena"; instead, look for where it explicitly states: "If X occurs, this framework fails." This is the only reliable method for determining whether it qualifies as physics. If you find that a core concept can always survive through parameter adjustment, can always retreat to an "explanation at the readout layer," and can never cause the framework as a whole to collapse, then your skepticism is justified, and you should reject the framework accordingly. This book places that authority of examination entirely in your hands and attempts to mark the failure point beside every major claim, so that you may exercise it.

The ideal mode of reading is recurrent rather than linear. When "closure" becomes clear in later chapters, return to the earlier diagnosis of container grammar; when "probability as readout residual" appears radical, return to the analysis of probability's unsupported status; when the minimal model appears crude, return to the principle of failure that runs throughout the book. The book's argument is not a straight line but a network that requires repeated cross-checking.

0.6. The Relationship Between This Book and Philosophy: Not an Excuse, but a Constraint

You will notice that this book makes extensive use of philosophical language: ontology, generation, process, settlement, and criteria of existence. Do not understand this as an excuse to replace physical calculation with philosophy. In this book, these concepts serve precisely the opposite function—they do not expand the scope of explanation but filter the acceptable starting points.

This point deserves to be stated fully, because it determines the legitimacy of philosophical language in this book. The book's position is that, if a mathematical structure is not generable at the ontological layer—that is, if it can only be assumed as a given background and cannot be attained as a generative result that may either succeed or fail—then, however technically elegant or computationally effective it may be, it should not be treated as a starting point of physics. Philosophy here exercises a form of veto power: it does not tell you which equations to write, but tells you which starting points are illegitimate. Terms such as ontology, generation, and settlement are therefore not rhetorical devices but constraints; each corresponds to an operational criterion—whether an event is generated, whether a structure closes, and whether closure freezes—and can therefore fail. A philosophical concept that cannot fail is no more permissible in this book than a mathematical premise that cannot fail.

0.7. If You Ultimately Reject This Book, That Is a Successful Outcome

This book was not written in order to be accepted. If, after reading it carefully, you conclude that generative logic cannot support scalable physics, that its closure criteria are so strict that they exclude the real world, or that the distinction between the readout layer and the generation layer is operationally unusable—and if you can identify precisely where the failure occurs—then this is exactly the outcome the book seeks to elicit. A theory that cannot be explicitly rejected is what is genuinely unphysical; the fact that you can identify precisely where "it breaks down" is itself evidence that it qualifies for physical consideration.

Thus, this book makes only one final request of the skeptic, a request far weaker than "please believe this book," yet far stricter: while reading, reject every concept that cannot fail—whether that concept comes from this book or from an existing theory to which you have long been accustomed. If you consistently do so, then, regardless of which side you ultimately take, you will already be standing on the side of physics. That, rather than agreement with the conclusions of this book, is what this preface truly asks you to bring into the main text.

Methodological Program

Before entering any specific argument, this part establishes the methodology followed throughout the book. It is not a conclusion but a constraint—a discipline that the book requires itself never to violate. It contains three elements: one principle governing all operations, a set of failure criteria that make the book refutable, and four propositions that the book explicitly assumes responsibility for and must justify independently.

Behind these three elements lies a single conviction, which the book states plainly at the outset: the qualification of a theory depends not on how much it can explain, but on whether it can clearly state where it ceases to hold. A theory that can always explain itself away—recasting every counterexample as "not yet occurred," a "low-probability event," or "outside the theoretical boundary"—deserves neither acceptance nor criticism or improvement, because it has already withdrawn from the testable domain. This book would rather be a framework that can be precisely rejected than an unfalsifiable worldview. All the disciplines below serve this objective.

0.1. The Difference Principle: Specification at the Ontological Layer, Operation at the Generation Layer

The entire argument of this book is built upon a distinction between levels. This distinction is stated in advance because every subsequent chapter relies on it, and many apparently difficult problems become clear once returned to it.

There are two levels that must not be conflated. The ontological layer answers the question "What is an event?" by specifying the event's intrinsic content. The generation layer answers the question "How do events compare and interact with one another so as to generate structures?" by specifying operations. The central claim of this book is that these two layers employ two different types of quantities that must not be confused—the ontological layer uses absolute quantities, whereas the generation layer operates only on differences.

More precisely, at the ontological layer, an event carries its intrinsic specifications; these specifications are absolute and constitutive. They form the existential content that makes the event the event it is, depend on no external reference frame, and require no comparison with anything else in order to hold. Yet these absolute intrinsic specifications are themselves inoperative—they enter no update rule. What enters the operations of the generation layer is never the absolute quantity of an event, but differences between events: the phase difference between two events, the difference between their rhythms, and the relative orientation between their directions. The absolute phase of an isolated event drives no evolution; what drives evolution is its phase difference from another event. The absolute direction of an event alone determines no coupling; coupling is determined by the relative orientation between two directional tendencies. Likewise, an absolute rhythm yields a judgment of whether it matches only when compared with another rhythm.

This is the Difference Principle: the ontological layer specifies an absolute triadic intrinsic structure, while the generation layer operates only through differences among such structures. Absolute quantities specify "what something is"; differences specify "how it is compared." The former are inoperative; only the latter are operative.

This principle is not a technical convention. It has three substantive consequences that run throughout the book. They are identified here and fulfilled individually below.

1. First, it realizes background independence through a mech-

anism. Background independence is usually expressed as a negative avoidance—"we do not use coordinates." This book transforms it into a positive constraint—"the generation layer may operate only on differences." The distinction is decisive: differences eliminate, by definition, any common external reference origin. A global translation of absolute phases or a global rotation of absolute directions automatically cancels out in operations on differences and therefore changes no judgment concerning coupling, closure, or freezing. This means that background independence is not a property cautiously maintained by the book after the fact, but a direct consequence of the construction that "the generation layer operates only on differences"—absolute quantities never enter operations, so there is no external reference on which those operations could depend. A formal verification of this invariance is given in Appendix B.

2. Second, it eliminates geometric circularity at its root. A pointed objection to any proposal for "generating space from relations" is that it uses direction to explain the generation of space, even though direction itself appears already to presuppose the space toward which it points. This is circular. The book's answer follows directly from the Difference Principle: direction enters generative judgments only through directional differences—the relative orientation between two directional tendencies—and relative orientation points to no spatial location. Even if coordinates are introduced for notational convenience in a numerical implementation, those coordinates are merely a bookkeeping device for differences and may be globally rotated arbitrarily without changing any judgment. Thus, "explaining space through direction" is not circular: what is used is always a background-free directional difference, not an absolute direction within a presupposed space.

3. Third, it establishes that logical steps belong to the generation layer. This book uses a discrete logical-step index to mark the progression of generation. A natural concern is whether this index covertly introduces time. The Difference Principle pro-

vides the answer: differences become operative quantities only when events encounter one another, and the sequence of "encounter—comparison—update" itself constitutes a beat; each logical step is precisely one advance of an operation on differences. Logical steps therefore count the number of advances made by operations on differences and intrinsically belong to the generation layer rather than the ontological layer—the ontological layer consists of specifications, which require no steps; the generation layer consists of advancing operations on differences, which necessarily proceed in steps.

The reader may treat this principle as the grammatical rule of the entire book: whenever operations at the generation layer are involved, only differences are permitted to appear; whenever an absolute quantity appears, it must belong to a specification at the ontological layer and must not participate in operations. The subsequent concepts—relation, closure, freezing, and readout—will all be shown to unfold this principle at different levels.

0.2. How to Falsify This Theory

This book does not claim necessity; it claims only conditional viability. This means that, if the world cannot be generated without presupposing a background, probability, or continuous time, and without relying on irreversible completion, then the book should be abandoned—no reinterpretation, correspondence principle, or future extension is intended to rescue it. A theory earns credibility not by explaining everything, but by specifying precisely where it can no longer stand.

For this reason, the book lists at the outset the ways in which it can fail. The following seven conditions are not exhaustive, but each is decisive: they are not rhetorical concessions but structural conditions under which the book loses physical meaning. Each is accompanied by a specific failure mechanism so that the reader may test it accordingly in the main text.

1. First, the failure of generative closure. This book assigns ontological meaning only to structures that achieve closure under internal criteria. If it can be shown that no nontrivial generative process can achieve closure at any scale, or that every apparent closure must depend on external truncation, a background structure, or observer intervention, then the central claim that "existence can be grounded in generative completion" collapses. In that case, "existence" must revert to an assumed primitive concept, and the book fails as an ontological framework.

2. Second, the failure of freezing as persistence. Closure alone is not sufficient. For a structure to count as a fact, it must freeze—it must persist as a unit reusable by subsequent generation without re-entering its own generative degrees of freedom. The book fails if closed structures cannot remain stable under any repeatable generative conditions; if persistence requires finely tuned initial conditions or external enforcement; or if no distinction can be drawn between transient closure and persistent fact. Without freezing, there is no mechanism for hierarchy, memory, or structural accumulation, and the world degenerates into fleeting events without factual continuity.

3. Third, the failure of irreversibility. This book derives the arrow of time from a non-negotiable condition: freezing is irreversible. The book fails if a frozen structure can be generatively "de-frozen" without losing its identity; if closure and freezing permit symmetric inverse operations; or if irreversibility must be reintroduced through statistical, thermodynamic, or boundary assumptions. In that case, the book loses its claim to provide a background-independent origin of temporal directionality—the arrow of time would once again become something requiring an external assumption rather than a trace left by generative completion itself.

4. Fourth, the failure of readout modulation. This book locates probability and statistical structure at the readout layer rather than the generation layer. This creates a strong and testable vulnerability and provides the book's most direct failure criterion:

if phase fidelity, resolution, or access to the generation layer is systematically improved, yet the observed statistical distribution does not change accordingly, then probability cannot be attributed to the limitations of readout; probabilistic structure must be ontic; and the book's readout-based explanation of quantum and statistical behavior fails. This is a definite point at which future experimental progress can directly falsify one of the book's central claims. Its operational form includes order-of-magnitude thresholds: measurable proxies such as coherence time, visibility, and the CHSH parameter should display calculable distributional drift when the readout protocol is changed under controlled conditions; if they do not, the book concedes its error.

5. Fifth, the failure of background independence. This book presupposes no spacetime, metric, continuum, or global parameter space. It therefore fails if generative rules cannot be formulated without an implicit embedding spacetime or state space; if closure conditions require continuous parameters, limits, or differentiable structures; or if the identity and comparability of events depend on prior coordinates or measures. Any unavoidable reintroduction of a background structure undermines the fundamental objective of the book. This criterion corresponds directly to the Difference Principle: the Difference Principle is precisely the construction intended to enable the book to pass this test.

6. Sixth, the failure of scalable structure. This book requires units of generation to be composable so that higher-level structures can form from frozen lower-level units. The book fails if closure cannot be nested across scales; if higher-level structures require fundamentally new generative principles; or if compositional hierarchies cannot emerge without external organizational rules. In those cases, the book cannot support a structured world extending beyond isolated facts—it could describe, at most, a collection of unrelated local closures, but could not explain how hierarchies form from quarks to hadrons and from hadrons to larger structures.

7. Seventh, the failure of conceptual economy. Finally, the book must remain conceptually minimal. It fails if additional primitive concepts must be introduced ad hoc to recover known phenomena; if closure, freezing, or readout proliferates into numerous unconstrained variants; or if the burden of explanation shifts from generative criteria to naming and reinterpretation. A framework that survives only through conceptual inflation has already failed its own methodological standard—it exchanges "one more mechanism" for greater coverage, and every additional mechanism that has not been unified represents a failure entered in the ledger.

The common feature of these seven criteria is that each points to a decidable and, in principle, accessible condition under which the book must concede that it is wrong. The book reserves no escape route by which a counterexample may be deferred indefinitely. When reading the main text, rather than first searching for what the book "explains," the reader should first search for the failure point of each core concept—if a concept can always survive through parameter adjustment, can always retreat to the readout layer, and can never bring down the framework as a whole, then skepticism toward it is justified.

0.3. Four Propositions Requiring Independent Justification

In addition to the preceding failure criteria, the book advances four propositions. They are neither self-evident nor mutually substitutable and must each independently bear the burden of justification. They are stated explicitly here so that the reader knows from the outset which argumentative debts the book incurs and where they will be repaid. Listing them separately is itself a discipline: it prevents the book from covertly employing an unjustified claim as though it were self-evident.

0.3.1. The No-Background Proposition: Structured Existence Can Be Achieved Without Presupposing Spacetime, Fields, or Any Continuous Background

This book treats this not as a postulate but as a claim that must be fulfilled chapter by chapter. This means that, whenever the book introduces a new concept—relation, weight, closure, or direction—it must demonstrate that the concept does not covertly depend on an ungenerated background. This proposition and the fifth failure criterion are two sides of the same requirement: the proposition states what the book seeks to achieve, while the criterion states how the book fails if it cannot achieve it. Part II, especially Chapters 5 and 6, is where this debt is primarily repaid: it shows that the intrinsic structure of events and relations between events can both be stated purely in terms of differences, which require no background.

0.3.2. The No Ontic Probability Proposition: Probability Is Not an Intrinsic Property of the World but a Consequence of Finite Readout Operations

The justification of this proposition cannot stop at the claim that "probability as ontology is unsettling." It must provide a specific mechanism through which probability arises at the readout layer and make that mechanism falsifiable. Chapter 3 first removes the legitimacy of probability as an ontological starting point through diagnosis; Chapter 18 provides the positive mechanism—how the triple truncation of readout through gating, bandwidth, and threshold compresses a determinate phase structure into a statistical distribution—and grounds its falsifiability in the fourth failure criterion. If this proposition fails, probability must be readmitted as a component of the generation layer, and the book's entire program concerning the readout layer collapses with it.

0.3.3. The No Unfailable Propositions Proposition: Any Physical Quantity or Criterion Treated as Fundamental Must Be Capable of Failure

There must be decidable conditions under which it does not hold. A "fundamental quantity" that can never fail lacks ontological qualification in generative terms. On the surface, this proposition serves as a critical tool directed at other theories—it is the common instrument used to diagnose the container, continuity, and probability—but the book applies it first to itself: every core criterion in the book—closure, freezing, and readout modulation—is required to state its own failure conditions, which gives rise to the seven preceding criteria. A theory that uses this proposition to criticize others while exempting itself would violate the proposition itself.

0.3.4. The Generation-Layer Accessibility Proposition: The Generation Layer Not Only Exists but Is at Least Partially Accessible

This is the easiest of the four propositions to overlook, yet it is also the most crucial, and therefore deserves full clarification here. It claims that structures at the generation layer can be traced and that their settlements can, in principle, be resolved—it is not a domain constitutively closed to every possible description.

Why must this proposition be justified independently rather than being covered by the preceding three? Because it addresses a different kind of question: not whether the generation layer exists, which belongs to the No-Background Proposition; nor whether probability belongs to the readout layer, which belongs to the No Ontic Probability Proposition; but whether our statements concerning the generation layer can possibly be true. Consider a position according to which the generation layer indeed exists and probability indeed arises at the readout layer, yet the generation layer is constitutively inaccessible to every description—readout is the only level at which anything is given to us, and we can never move beyond readout to reach the generative structure that it truncates. Under such a position, all the book's claims concerning the readout layer—"probability arises from readout truncation" and "the measurement problem is dissolved at the generation layer rather than displaced within the readout layer"—might be ontologically true, yet could not be epistemically fulfilled: we would be unable to specify what the "prior structure" truncated by readout actually is or explain why it is that structure.

The book therefore openly acknowledges a dependency: the strength of its claims concerning the readout layer is exactly equal to the strength of the Accessibility Proposition. This is not a weakness but an act of honesty—the book makes its epistemological premise explicit rather than leaving it concealed behind the argument. More importantly, the Accessibility Proposition is itself falsifiable rather than a secure metaphysical assumption: its empirical stake is precisely the fourth failure criterion. If changing the readout protocol produces no calculable drift in the statistical distribution, then not only does the claim that "probability arises from truncation" fail, but the Accessibility Proposition is simultaneously falsified—because this would mean that no channel exists between readout and the generation layer that can be manipulated and resolved by us. The Accessibility Proposition and the No Ontic Probability Proposition therefore share responsibility: they are not two independent wagers, but two sides of the same wager, jointly tested by the same set of experiments.

The methodological program is now established: one principle—the Difference Principle; seven criteria—how the framework can fail; and four propositions—the debts it incurs. All subsequent work in the book may be read as a gradual answer, under these constraints, to the same question: how structured existence can be achieved without presupposing a background, probability, or continuous time, and where that achievement may fail.

Introduction

0.1. Background Independence Is Not a Technical Choice but an Ontological Choice

This book begins with a question that contemporary theoretical physics commonly treats as a "technical premise": must every physical theory take a continuous, differentiable spacetime structure as its background? On this question, the book adopts a clear and unavoidable position—background independence is not a modeling preference or computational strategy, but an ontological choice. It determines not only what a theory permits as a starting point, but also what it assumes already exists before generation.

Within the mainstream frameworks of modern physics—whether classical field theory, general relativity, or quantum field theory—substantial formal differences often conceal a shared assumption: a continuous, differentiable spacetime structure is given in advance. Physical quantities are defined as functions on this structure, dynamics are expressed through differential equations on it, and change is understood as evolution along continuous parameters. Even in the most revolutionary theories, this assumption is usually not discussed explicitly but is treated as a "neutral background" whose validity is self-evident.

Yet this "neutrality" exists only at the technical level. At the ontological layer, a differentiable manifold already carries highly nontrivial commitments: continuity, local comparability, infinite divisibility, and presupposed relations of adjacency. Once these structures are accepted, the problem of generation has already been answered prematurely at the starting point—the object of discussion is no longer "how structure appears," but "what occurs upon structure." This is not an innocuous simplification. It logically conceals what genuinely requires explanation: where the continuous background treated as a stage itself comes from.

The most common defense is that, even if a continuous background is not ultimately real, it remains an extraordinarily successful effective description, and that more precise equations, higher-order corrections, or more complex symmetries may be sufficient to resolve existing difficulties. This defense overlooks a fundamental distinction: precision cannot repair a starting point. Within a framework that has already assumed continuity, every formal improvement can operate only within that assumption; it may fit curves more accurately, but it cannot explain where continuity itself comes from or why concepts such as "change," "interaction," and "simultaneity" possess precisely their present forms. Improving precision and auditing a starting point are tasks of different kinds; the latter cannot be accomplished by the former.

This book therefore takes a different path: it begins not with space, time, fields, or probability, but with events; it assumes no geometry or continuous parameter and permits only discrete generative steps, relation updates, and closure criteria. At this logical position, gradients, densities, manifolds, and differential operators have not yet been introduced; whether and how they appear must be explained as subsequent results rather than assumed as premises.

It must be emphasized that this position does not deny the effectiveness of existing theories. The book does not seek to modify or replace any particular continuous field theory. Its concern lies at an earlier level: under what conditions do such descriptions acquire logical legitimacy? Every theory formulated on a differentiable manifold must itself depend on some more fundamental generative conditions for its validity; this book is a systematic investigation of those conditions. In this sense, background independence is neither the technical operation of "removing coordinates" nor the formal adjustment of "adopting a different mathematical language," but a redefinition of the ontological starting point of physical theory. Once this redefinition is accepted, the appearance of simultaneity, probability, stable structures, and even "facts themselves" can be understood uniformly as results of generation—closure—freezing, rather than as properties of a presupposed stage.

This book begins before the background. That is its first and deepest disagreement with almost all existing physics.

0.2. The Position of This Book Within the EQT Framework

This book belongs to the broader research series of Energy Quantum Theory (EQT), but occupies within it a specific position that is easily misunderstood. Clarifying this position is a prerequisite for reading the book correctly—especially for understanding which tasks the book neither assumes nor attempts to assume.

First, this book does not model specific physical interactions. The strong, weak, electromagnetic, and gravitational interactions, together with phenomenological explanations of dark matter and dark energy, all lie outside its scope. Nor does the book directly fit any experimental data or provide numerical predictions that can immediately be compared with observations. Such work has either been undertaken in other EQT volumes or should be deferred until the generative framework itself has become stable. Pursuing phenomenological precision before the generative starting point has been clarified would repeat the error diagnosed by this book: concealing an unaudited starting point beneath a more precise description.

Under this premise, the subject of the book is deliberately narrowed to four terms: generation, closure, freezing, and failure. Generation concerns how events and relations appear and evolve under background-free conditions; closure concerns which generative processes can attain internal consistency and form decidable structures;

freezing concerns the conditions under which such structures acquire persistence and enter the level of facts; failure concerns situations in which closure cannot be attained, structures cannot be sustained, or generation is interrupted. The book repeatedly emphasizes a position that distinguishes it from almost all existing theories: failure is not an anomaly, but an output as important as success. A framework capable only of producing success and incapable of producing failure has covertly written existence into its premises.

Within this restriction, the book does not answer the question "Which forces constitute the universe?" but an earlier question: under what conditions can the concept of "force" become possible? It does not discuss "what a particle is," but asks "how stable objects are possible." Readers searching for familiar field equations, symmetry groups, or parameter fitting will therefore inevitably be disappointed—this disappointment is a deliberately established boundary effect of the book, not an omission.

This position directly determines the relationship between this book and the preceding nine EQT works, and that relationship must be made clear to avoid misunderstanding. The first nine works are built upon the language of differentiable manifolds and continuous fields. Their objective is to reveal, within existing frameworks of physical description, the unifying explanatory power offered by frequency, gradients, and nonlinear cascades; their concern is "how physics can be reinterpreted against a continuous background." The starting point of this book is entirely different: it does not assume the legitimacy of the continuous background itself. Accordingly, this book is neither a summary of the preceding nine works nor their direct extension, but is situated logically at an earlier level: the continuous variables, gradient descriptions, energy densities, and flow equations employed in those works must all be reinterpreted here as effective readouts formed after generation. If the first nine works answer "how physics can be rewritten on a manifold," this book answers "why manifold-based descriptions are usable under certain conditions."

This relationship has a strict consequence: the conclusions of the preceding nine works cannot be treated as premises of this book. Here, they may serve at most as potential downstream applications or objects of consistency checks, rather than as logical starting points. If the arguments of this book cannot stand independently, then even formal consistency with the preceding nine works provides no theoretical justification; conversely, even if the book cannot recover all their physical content in the short term, its generative logic must still be assessed independently. This book is therefore not the completed manuscript of a "unified theory," but a generative manual for a unified theory: it does not answer "Which equations does nature obey?" but "What kinds of generative structures make equations meaningful descriptions?" What it provides for EQT is an ontological starting point, not a phenomenological endpoint.

0.3. The Self-Imposed Constraints of This Book

The position described above requires the book to impose strict constraints upon itself. These constraints were formally stated in Part Zero as the Difference Principle, seven failure criteria, and four propositions requiring independent justification. Their technical content will not be repeated here; instead, only their function as constraints will be identified, because these constraints determine which questions the book is subsequently permitted to ask and which answers it is permitted to accept.

The constraints operate at three levels. At the ontological starting point, the book does not permit spacetime, objects, fields, or states to serve as starting points of generation—all such structures may appear only as readouts after generation has been completed and frozen; the event is the only permitted minimal ontological unit. At the level of generation and failure, the book requires every fundamental structure to be capable of failure, and failed structures are not retained as "potential states" or an "inventory of unrealized possibilities"—the book rejects semantics involving sample spaces, possible worlds, or complete sets of hidden states, because they are disguised forms of "never failing." At the level of criteria, the book requires existence to be premised on decidable closure, closure to be achieved within a finite number of generative steps, freezing to be irreversible, and time, causality, and probability all to belong to the readout layer—any attempt to replace finite criteria with infinite limits, asymptotic convergence, or "arbitrarily close approximation" is treated as incomplete.

These constraints hold as a whole and cannot be partially abandoned while the remainder are retained. Abandoning any one of them would cause the theory to slide back into the container grammar of "object—time—state"—precisely the place this book seeks to leave. Their function is therefore not to generate conclusions directly, but to delimit the legitimacy of explanations and prevent conceptual regression: questions extending beyond this boundary are treated as undefined within the framework of the book.

Under these restricted conditions, the task of the book is not to construct a grand explanatory system, but to test one issue: whether generation, structure, and stability remain possible under such stringent constraints. If they are possible, subsequent physical theories acquire a new ontological foundation; if they are not, that failure itself constitutes an important counterexample to the intuitions underlying existing theories. Whatever the outcome, this book will constitute a genuine starting point that can be questioned, tested, and abandoned—this is all that it requires of itself.

Part I.

Diagnosis: Why Existing Physics Must Be Suspended

Introduction

The Introduction established the position of this book: to begin before the background and not to take continuous spacetime, probability, or unfailing limits as the starting point of generation. Yet a position, however clearly stated, is not itself an argument. Before proposing any positive generative framework, this book must first answer a prior question: why is the existing language of physics insufficient to address the problem of "generation"? Unless this question is answered first, "beginning before the background" will sound merely like a posture rather than a compelled conclusion. The task of Part I is to transform this posture into a conclusion—through diagnosis, by showing that the starting points of existing physics themselves prevent the problem of generation from being posed.

The nature of "diagnosis" must first be clarified, because it determines how this part should be read. Diagnosis is not refutation. This book does not seek to prove that relativity or quantum mechanics is wrong—they have been repeatedly tested within their respective domains of applicability and are successful theories, which this book fully acknowledges. The diagnosis is directed not at their empirical content, but at the grammatical premises they share and yet almost never examine as assumptions. A theory can be extraordinarily precise within its container while remaining entirely silent about where that container came from. The object of diagnosis is precisely this silence: those things silently treated as starting points and thereby exempted from all examination. Part I therefore does not compete with any specific theory, but steps back to examine the ground on which they all stand.

This book identifies three such premises. They have been inherited so naturally that they are usually regarded not as premises, but as the only possible way to think about physical questions.

The first is the container (Chapter 1). Almost all modern physics adopts a structure that this book calls "container grammar": first provide a background that contains everything, then place content within it, and finally allow that content to evolve along some parameter. From Newtonian absolute space to Minkowski spacetime, from the dynamical metric of general relativity to the Hilbert space of quantum mechanics, the geometry of the container has been repeatedly transformed, but the structure in which "the container precedes the content, and the content is placed upon it" has never been abandoned. Chapter 1 will show that, even in theories with the strongest form of background independence, the geometry of spacetime is dynamical, but the existence of spacetime is not—it never needs to be generated and can never fail. This is precisely where the problem of generation is covertly answered at the starting point.

The second is continuity (Chapter 2). The long-standing stability of container grammar depends on a deeper and less frequently questioned assumption: continuity. Continuous variables, continuous evolution, and continuous limits are treated as the "natural language" of physics, while discreteness is instead required to demonstrate how it can "recover continuity." Chapter 2 will show that continuity is not a neutral choice of representation, but an active ontological commitment—it quietly replaces the question "why does this state exist?" with "how did this state evolve from the preceding state?" and thereby makes the generative question of emergence from nonexistence impossible to pose. More importantly, the continuous limit functions as a failure absorber: every mismatch can be deferred to a "finer scale," making the theory almost incapable of conceptual failure. A premise that cannot fail does not, by the standards of this book, qualify as physical.

The third is probability (Chapter 3). Probability is ubiquitous in modern physics, yet it has never been defined at the generation layer. It is calculated, manipulated, and fitted, but never appears as something that can be generated or can fail; it is always appended from outside the dynamics through an additional rule. Chapter 3 will show that probability appears only after measurement, varies with the mode of measurement, and can be modulated by measurement conditions. These three indications jointly support one judgment: probability is not an intrinsic property of the world, but belongs to the readout layer. Once ontologized, probability becomes immune to failure: every deviation can be attributed to fluctuation, and every counterexample can be deferred to a larger sample. This unfalsifiability is not an advantage, but a sign of misplacement—it indicates that probability has been assigned to the wrong layer.

These three premises—the container, continuity, and probability—arc not isolated from one another. They are three aspects of a single mode of thought that mutually support one another: the container requires continuity to fill it, continuity requires probability to manage the mismatches it cannot explain, and probability in turn requires the container to provide the sample space and moments on which it depends for definition. Part I dismantles them one by one, not to produce three separate criticisms, but to clear a single position: a position that presupposes no container, no continuity, and no probability. At this position, "generation" can no longer be discussed as some process within a container, but can instead be posed as something prior to the container that enables the container itself to appear as a readout.

It should be stated in advance that all the work in this part is negative. It establishes nothing; it only clears away. After the three diagnostic chapters, the book will not immediately possess a theory of generation; it will possess only an emptied starting point. Yet this clearing is necessary: every concept introduced in the positive construction beginning in Part II—event, relation, closure, and freezing—must be able to return and pass the tests of Part I by demonstrating that it does not covertly depend on any of the three premises that have been removed. In this sense, Part I is not only the beginning of the book, but also a line of examination running throughout it: whenever a concept is found to reintroduce a container, continuity, or ontic probability, it fails along this line.

1. The Hidden Postulate of the Container Universe

In modern physics, the least questioned yet most consequential statement is: "Events occur in spacetime." On the surface, this appears to be merely descriptive language. In practice, however, it simultaneously performs three roles without being noticed: it is an ontological premise, according to which spacetime exists prior to events; a formal structure, according to which events are defined as points or regions in spacetime; and a direction of explanation, according to which the properties of spacetime explain events, while events in turn explain spacetime. It is precisely the combination of these three roles that has long shielded the statement from systematic examination and gradually caused it to be mistaken for a logically neutral starting point.

This chapter reveals that this starting point is not neutral. Once "events occur in spacetime" is adopted as the point of departure of a theory, physical explanation inevitably falls into a circle—it uses an ungenerated container to define events and then uses events to explain that container, while the true starting point remains absent. The diagnosis proceeds in seven steps: from how descriptive language becomes an ontological premise (1.1), through the circular structure of formalization (1.2), to how continuity stabilizes this circle (1.3), and then to how this premise becomes successively institutionalized in cross-theoretical container grammar (1.4), axiomatization (1.5), numerical limits (1.6), and probabilistic averaging (1.7), before becoming concentratedly visible in three well-known predicaments (1.8). The objective of this chapter is not to deny the effectiveness of spacetime descriptions, but to show that this effectiveness comes at a cost, and that this cost falls precisely upon the problem of generation.

1.1. How Descriptive Language Quietly Becomes an Ontological Premise

In ordinary physical discourse, an "event" is usually understood as a process occurring at a particular location and at a particular time. A particle scattering is described as "occurring at a spatial point during a temporal interval"; a measurement is described as "producing a result at the location of the apparatus at a certain moment." In this context, spacetime appears to be merely a coordinate system for recording information—a neutral and passive tool of notation. If spacetime were nothing more than this, it would raise no ontological problem.

Yet once a theory enters the stage of formalization, this descriptive language undergoes a quiet change of role. Events are no longer merely "marked by coordinates," but are defined as points, regions, or trajectories in spacetime. From that moment onward, spacetime is no longer a recording tool, but a precondition for the existence of events: without spacetime, there is no position for an event to occupy, and therefore no event. A recording tool and a precondition of existence occupy entirely different ontological positions, yet physical formalization replaces the former with the latter without declaration.

This substitution is usually not consciously recognized. Instead, it becomes institutionalized as something "self-evident." It relies on several apparently irrefutable questions: if events have no positions, how can they be distinguished? If events have no time, how can they be ordered? If events are not embedded in a continuous structure, how can equations describing them be written? The force of these questions lies in the fact that they are all true—within container grammar, positions are indeed required for distinction, time is required for ordering, and continuous structures are required for writing equations. Yet their truth rests precisely on the prior acceptance of the container premise. They are not proofs that "spacetime precedes events," but echoes of that premise. To defend the ontological priority of spacetime through these questions is to use the conclusion to prove the premise.

The position that "spacetime precedes events" has therefore never actually been demonstrated. It has been accepted by default, driven by a set of circular questions and without justification. The first step of diagnosis is to relabel this default as what it originally is: an untested ontological commitment, rather than a neutral convention of notation.

1.2. The Circular Structure of Formalization

Once spacetime is established as the container of events, the formalization of a theory immediately enters a closed circle. The typical path of this circle can be stated clearly: first, assume a spacetime structure—a continuous manifold, dimensionality, a metric, or their discrete variants; second, define events upon this structure—as points, curves, or local excitations of fields; third, use the evolution of events to explain the properties of spacetime—curvature, causal structure, or dynamical geometry; and finally, use the resulting spacetime structure to explain why events evolve as they do.

Within this circle, spacetime is both premise and result, both an explanatory instrument and an object to be explained. General relativity is the most refined embodiment of this circle: the distribution of matter determines the curvature of spacetime, while the curvature of spacetime in turn determines the motion of matter. This mutually determining structure is mathematically self-consistent and empirically successful, and precisely for this reason its circularity is rarely noticed. Yet self-consistency is not the same as possessing a starting point. A circle whose beginning and end are joined, however rigorous, has no point of departure independent of the other links; it explains the relations among the components within the circle, but does not explain why the circle itself exists.

The cost of this circle is most clearly exposed when the theory encounters extreme conditions. When confronted with singularities, problems of initial conditions, or the rupture of quantum measurement, the standard response is that "spacetime fails here" or that "spacetime requires quantum correction here." It is worth noting that such responses do not break the circle, but merely push it to the boundary: they assume that spacetime ought to hold everywhere and that its failure is only a local anomaly awaiting repair. But if spacetime is from the outset a product of the circle rather than an independent starting point, then its "failure" in certain places should not be treated as a defect requiring repair, but as evidence of the limitations of the circle itself. This book will return to this point later (see the discussion of singularities in Section 1.8); for now, it is sufficient to note that pushing failure to the boundary is a way of maintaining the circle, not resolving it.

1.3. Continuity Is Not a Neutral Assumption

Container grammar does not occupy the starting point by itself. Its stability depends on a deeper and less frequently questioned premise: continuity. Whether in classical mechanics, field theory, general relativity, or the mainstream formulations of quantum theory, continuous variables, continuous evolution, and continuous limits are introduced naturally and treated as convenient tools of mathematical expression. Yet this "convenience" is not value-neutral. Once continuity is adopted, it systematically shapes which questions may be posed, which explanations are acceptable, and how failure is handled. The central argument of this section is that continuity is not a passive choice of representation, but an active ontological commitment—it quietly determines how events are permitted to occur, how they may be compared, and how they may be explained, and thereby provides the stabilizing mechanism for the circle described in the preceding section.

The status of continuity in physics derives from its immense success in classical theories: Newtonian mechanics describes trajectories using a continuous time parameter, electromagnetic phenomena are understood as continuously distributed field quantities, and the thermodynamic limit produces macroscopic stability through continuous variables. These successes gradually caused continuity to be regarded as the "natural language" of physics, while discreteness came to be treated as a special case requiring additional justification. This historical path produced an important consequence: continuity ceased to be regarded as an assumption and instead became the default minimum standard of rational expression—any discontinuous model is often required to demonstrate how it "recovers continuity" in some limit, while continuous models are rarely required to explain where their continuity comes from.

The most direct effect of continuity is to dissolve the problem of the starting point. Within a continuous-time framework, every state is understood as the result of an infinitesimal evolution from a preceding state; consequently, "why does this state exist?" is naturally replaced by "how did this state evolve from the preceding state?" The problem of generation is thereby rewritten as a problem of evolution. This substitution is formally legitimate, but ontologically costly: it assumes that states have always already existed and merely continue to change; it does not permit the question of emergence from nonexistence to be posed; and it interprets failure as a poorly taken limit rather than a mechanism that failed to hold. Within a continuous framework, events never truly "appear"; they are merely points on already given trajectories.

Continuity also provides an exceptionally effective buffering mechanism against failure. Almost any mismatch, instability, or nonconvergent behavior can be interpreted as an inadequate approximation, insufficient resolution, an excessively large numerical step, or a limit not yet reached. This explanatory route is extremely successful in engineering and computation, but it has a pronounced anesthetizing effect on generative logic: so long as the continuous limit is retained, the theory can almost never fail at the conceptual level—the problem can always be deferred to a "finer scale." Continuity is therefore not merely a mathematical structure, but also a mechanism of theoretical immunity.

This immunity is further strengthened by the binding of continuity to differentiability. In modern physics, states are required not only to be continuous, but also differentiable; evolution is not only continuous, but governed by differential equations. This binding imposes a further constraint: permitted changes must be smooth, genuine jumps or settlements are excluded, and stability is understood as invariance under infinitesimal perturbations. Yet the problem of generation concerns precisely those forms of structural formation that are nonsmooth, non-gradual, and nonlinearly triggered. When differentiability is established as the default condition of legitimacy, such questions are excluded before they can enter the theory.

A common defense holds that continuous models are more general than discrete models, and that the latter are merely special cases or approximations of the former. This claim confuses two meanings of "more general." In the mathematical sense, continuous structures can indeed accommodate richer classes of functions; but in the generative sense, continuity instead excludes certain possibilities—it excludes settlement as an event, excludes the foundational status of thresholds and gating, and excludes failure as a structural result. Continuity is therefore not "more permissive," but recuts the space of questions. This is especially evident with respect to temporal direction: continuous time understands time as a uniformly flowing parameter that can be divided without limit. Under this assumption, the arrow of time must either be added externally, as in thermodynamic assumptions, or explained through an asymmetry of initial conditions; irreversibility is never genuinely generated, but merely labeled. Continuous time is well suited to describing evolution, but poorly suited to explaining why evolution remains in certain structures.

For this reason, this book does not seek to replace existing theories with a "better continuous model," nor does it attempt to repair the problem of generation through discretization techniques within a continuous framework. Instead, it adopts a more radical but logically clearer position: at the generation layer, continuity should be suspended in its entirety—evolution should not be indexed by continuous parameters, stability should not be explained through limiting processes, and smoothness should not serve as a precondition of legitimacy. Only under this suspension can generation, closure, freezing, and failure be introduced as first-order concepts rather than treated as subsidiary phenomena of continuous evolution. Continuity has long escaped systematic questioning not because it is irreplaceable, but because it is exceptionally successful, exceptionally convenient, and exceptionally effective at absorbing failure. Yet it is not an isolated assumption, but is embedded within a more general theoretical grammar—a grammar that belongs to no single theory, but is jointly inherited across different physical paradigms. The next section examines this grammar.

1.4. From Newton to Quantum Field Theory: The Container Has Never Been Generated

When this diagnosis is situated within the history of physics, container grammar can be seen to persist under different disguises, while its core remains unchanged.

In Newtonian mechanics, space and time are absolute, rigid, and content-independent containers. Objects move within them, but the containers themselves participate in no physical process; they passively provide positions and moments. This is the purest and most transparent form of container grammar—the stage and the actors are entirely separate, and the stage exists first.

Special relativity profoundly transformed this container: space and time were no longer independent, but were unified into Minkowski spacetime, and different observers no longer agreed on judgments of "simultaneity." This was the first major revision of absolute spacetime in physics. Yet it did not alter the core of container grammar: Minkowski spacetime remained a prior and rigid stage, upon which events were still placed and assigned coordinates. What changed was the geometry of the stage, not the structure in which "the stage precedes the content."

General relativity went further by making the container itself dynamical: the geometry of spacetime is determined by matter and energy and in turn governs the motion of matter. For this reason, general relativity is often said to have achieved background independence. A precise qualification is required here, and it is central to the diagnosis of this chapter: in general relativity, the geometry of spacetime is dynamical, but the existence of spacetime is not. The metric field may evolve, curve, and carry gravitational waves, but it always remains an already existing, continuous, four-dimensional manifold; it never needs to be generated and can never fail—there is no result in which "spacetime failed to form here." Einstein's equations describe how spacetime curves, but never why spacetime exists. Background independence here reaches only the level of "geometric plasticity," not the level at which "existence can be generated and can fail."

Quantum theory is often regarded as the most radical overthrow of the classical conception of reality. Yet with respect to the container, it not only fails to escape container grammar but continues it in a more concealed form. Quantum mechanics retains a prior stage: Hilbert space, together with unitary evolution governed by a Hamiltonian, and an externally added spacetime background within which measurements occur at particular places and times. Two remnants of container grammar are especially noteworthy. First, Hilbert space, as the container of states, has a dimension and structure given a priori rather than emerging from any more fundamental process. Second, and more importantly, measurement is formulated as a "collapse occurring at a certain place and time"—which presupposes an already existing spacetime point, with collapse merely appended as an additional rule upon it. This is one structural source of the persistent measurement problem: collapse is required to occur on an already given stage, and therefore can exist only as an externally added postulate rather than being understood as a result of generation.

Quantum field theory reconceives particles as excitations of fields and therefore appears to move beyond the particle picture. Yet fields are presupposed as prior existents extending throughout spacetime, and excitations are merely local behaviors of fields against a given background; events of creation and annihilation still occur within an ungenerated field–spacetime container. The bearer of ontology changes from particles to fields, but the grammar in which "objects precede generation and content is placed within a background" remains entirely unchanged.

From Newtonian mechanics to quantum field theory, the form of the container is repeatedly renewed—absolute space, Minkowski spacetime, dynamical metrics, Hilbert space, and quantum fields—but one feature persists throughout: the container has never been generated. Its geometry may be incorporated into dynamics, but its existence always remains a premise. This is precisely the position this book seeks to leave.

1.5. How Axiomatization Conceals the Generative Gap

In modern physical theories, axiomatization is often regarded as the highest form of rigor. If a theory can be expressed as a clear set of axioms, definitions, and consequences, it is frequently considered logically "well founded." Yet this criterion of evaluation does not address a more fundamental question: do these axioms genuinely ground the objects under discussion, or do they merely fix their operational form? The central claim of this section is that axiomatization in mainstream physical theories does not answer the problem of generation, but removes the generative gap from the space of questions through formal closure.

Axiomatization originally served as a technical method intended to reduce implicit assumptions, clarify inferential paths, and ensure internal consistency. In mathematics, this objective is reasonable and sufficient. Yet when axiomatization is directly transferred into physical ontology, a problem arises: consistency is mistaken for the source of existence. Physical axioms often begin with the phrase "given a certain structure"—given spacetime, given a state space, given an evolution operator. From such a starting point, the task of the theory is no longer to explain how these structures are generated, but to explain how they operate within their own framework; the problem of generation is thereby excluded at the outset.

Logically, "being given" and "being generated" belong to entirely different categories. A given object is assumed already to exist, and its legitimacy need not be explained within the theory; the existence of a generated object must be produced by a mechanism, and the process of production itself can fail. Axiomatization naturally inclines toward the former: by explicitly declaring that "a certain object exists," it transforms the object's existence into a logical premise rather than a dynamical result—as long as the axiomatic system is consistent, the object is regarded as legitimate. Generative ontology requires precisely the opposite: existence is not declared, but achieved.

This tendency produces an important consequence: an axiomatic system is logically closed. Once its axioms are accepted, any failure within the system is usually interpreted as an error of derivation, an error of calculation, or an inappropriate choice of boundary conditions or parameters. A fourth explanation almost never appears—that the axioms themselves failed to generate the object they claim. Failure can occur only within the process of reasoning, not at the level of existence; the existence of the object is fixed in advance and therefore does not permit a state of "not generated" to be identified as a theoretical failure. Correspondingly, time in axiomatic theories usually appears only as a parameter or index. Its role is limited to ordering states and parameterizing evolution, rather than serving a generative function; a state at any position on the temporal axis is regarded as legitimate so long as it satisfies the axioms. This produces a crucial consequence: there is no state that has "not yet become a fact"—every state either exists or is excluded for failing to satisfy the axioms, rather than because it has not yet been achieved. The incompleteness distinctive of generation has no place within such a temporal structure.

Axiomatization also systematically dissolves thresholds and gating. Generative processes often depend on thresholds: only when conditions are satisfied to a certain degree does a structure form; when conditions are insufficient, generation fails. Axiomatic systems, however, usually avoid introducing genuine thresholds. Even when thresholds appear, they are often represented as parameters within continuous functions and can be approached without limit. This treatment transforms gating mechanisms into problems of continuous parameter adjustment, thereby avoiding the decisive moment at which a structure either forms or does not form. Generation is redescribed as gradual change, and settlement is diluted into limiting behavior. This reveals a long-neglected distinction: axiomatic consistency is not equivalent to generative sufficiency. A theory may be logically complete yet generatively empty—it may rigorously characterize the properties of a class of structures while failing to explain why those structures appear, why they remain stable, or why they do not arbitrarily collapse. Axioms state "if it exists, then this is how it behaves," but do not answer "why does it exist?"

Ultimately, failure becomes invisible within an axiomatic system. Generative ontology requires failure to be an expressible, analyzable, and distinguishable result; but within an axiomatic system, if a configuration does not satisfy the axioms, it simply lies outside the theory's domain of discussion—it is not a "result of generative failure," but merely an "illegitimate object." This treatment is logically clean but physically problematic: nature does not cease attempting to generate structures merely because they are "illegitimate." Failure is part of the process, not an excluded anomaly. This book therefore does not deny the value of axiomatization, but clearly defines its boundary of applicability: axiomatization is suitable for describing the internal relations of frozen structures, but not for grounding generation itself. The strategy of this book is not to "propose a new axiomatic system," but to introduce generative rules capable of failure—rules that do not guarantee success, do not presuppose existence, and do not acquire legitimacy through consistency. Their only criteria of evaluation are whether they can generate, under what conditions they generate, and where and why they fail. When existence is written into axioms, generation is removed from the theory; and once responsibility for generation has been removed, the next step is no longer logical closure but practical buffering—numerical limits assume control of failure precisely at this point.

1.6. Numerical Limits and the Myth of "Not Yet Converged"

In contemporary physical practice, numerical computation is no longer merely an auxiliary tool of theory, but an important component in the production of facts. Many physical conclusions do not arise from analytic solutions, but from simulation, iteration, approximation, and statistical averaging. Within this practical context, an apparently neutral judgment appears repeatedly: "the result has not yet converged." The central claim of this section is that "not yet converged" is not a purely technical description, but a narrative device with definite ontological consequences—in practice, it systematically rewrites generative failure as incomplete computation, thereby delaying or even eliminating accountability for the theory itself.

In the era dominated by analytic solutions, whether an equation could be solved was itself part of a theory's capacity: insolubility meant that the theory failed on the problem, or at least indicated that its existing structure was insufficient to support the description. The introduction of numerical methods altered this situation—through discretization, iteration, and approximation, a theory no longer needed to provide a closed-form solution, but only a computable update rule. Whether the result was "correct" was transformed into whether a stable output could be reached within the limits of available computational power. In this transition, a crucial semantic change occurred: failure was no longer expressed as "the theory cannot generate," but rewritten as "the computation is not yet sufficient."

At the same time, numerical computation introduced a new criterion of legitimacy: convergence. If the result tends toward stability as the step size is reduced, the grid is refined, or the sample increases, it is regarded as "credible"; if the result is unstable, the instability is attributed to insufficient parameters, truncation error, or limits of computational power. This criterion is technically reasonable, but its ontological meaning is often ignored: the judgment of existence is outsourced to a limiting process. A structure no longer needs to be generated in finite steps; it only needs to "be expected to appear" in some idealized limit, and that limit can always be deferred in practice. Thus, what should be an explicit failure from the perspective of generative ontology—insufficient conditions, a mismatch of mechanisms, unmet gating, or the nonformation of structure—is uniformly rewritten in numerical narratives as an insufficiently small time step, inadequate sampling, poor initial values, the need for higher precision, or longer evolution. Failure is thereby de-ontologized: it is no longer "not generated," but merely "not yet computed." This rewriting has a significant protective effect: no matter how unstable the results may be, the theory itself can temporarily be exempted from responsibility.

A common expression in numerical practice is: "in principle, as the step size tends to zero and the sample tends to infinity, the result should converge to a certain value." This statement is mathematically unobjectionable, but in physical ontology it introduces a dangerous assumption: existence can be deferred to an unreachable limit. Generation no longer needs to be completed in the finite world; it only needs to hold under imagined infinite precision. "Existence in principle" gradually replaces "generation in actuality." A common categorical slippage is also hidden here: numerical stability is not equivalent to generative closure. A numerical process may be stable, but this does not mean that it has generated a physical fact—stability may arise from numerical dissipation, algorithmically built-in smoothing mechanisms, averaging operations, or the choice of artificial thresholds. Such mechanisms can suppress divergence, but they are not equivalent to the generation of a structure with internal closure.

Numerical simulation also has a structural tendency: it always permits "one more step." When a result is unsatisfactory, the evolution time can be extended, the grid refined, parameters adjusted, or random seeds changed. This extensibility is an advantage in engineering practice, but from the perspective of generative ontology it is problematic—because it dissolves the point of judgment at which one must say "it fails here." Failure is no longer an event, but a possibility deferred without limit. In the final analysis, the phrase "not yet converged" implicitly contains at least three unstated postulates: the sought structure exists in principle; failure arises from external technical limitations rather than an internal generative mechanism; and sufficient computational power can replace the generative capacity of the theory itself. Together, these postulates form a kind of computation-centered ontology—the view that existence can be approached through the accumulation of computational resources.

This book therefore does not deny numerical methods, but requires their role to be redefined: in generative ontology, the task of numerical computation is not to "approximate an assumed existing solution," but to test whether generative rules can achieve closure in finite steps. The key question is no longer whether a process converges, but whether a definite moment of settlement appears, whether there is an uncrossable gating threshold, and whether "not yet generated" can be distinguished from "never generated." This requires a stricter standard: a theory must assume the risk of failure under finite conditions—if a structure can hold "in principle" only under infinite precision, infinite time, or infinite computational power, then it does not hold in the generative sense. This is not technical severity, but ontological honesty. "Not yet converged" is appealing because it gives the theory an almost unlimited buffer, but it is precisely this buffer that conceals the generative gap, delays failure, and virtualizes existence. Yet this is not the endpoint at which failure is dissolved: when numerical instability repeatedly appears, theory still has a more radical form of rewriting available—handing the problem of generation as a whole over to probability.

1.7. Probabilistic Averaging: The Final Step in the Statisticalization of Failure

If the continuous limit provides a theory with an escape route of "existence in principle," and numerical approximation provides failure with the buffer of "not yet completed," then probabilistic averaging completes the final step: it rewrites generative failure as a statistical distribution. Once this rewriting is complete, failure no longer appears in any explicit form—it is not marked, not reported, and not even discussed, but is systematically absorbed into the language of probability.

In the sense of generative ontology, a structure is either generated or not generated, and closure either occurs or fails. This is an event-type judgment. Probability introduces another paradigm: instead of asking whether a single act of generation holds, it turns to the statistical distribution of repeated experiments; thus, the question that should have been asked—"why was it not generated?"—is replaced by "what is the probability of this result?" The problem of generation is shifted as a whole into a problem of distribution. A notable feature of probabilistic description is that it permits predictive power to be retained without specifying the generative mechanism—as long as the form of the distribution agrees with experimental results, the theory can temporarily avoid questions such as why this result appeared, why another result did not appear, and why structure failed here but succeeded there. The absence of mechanism is smoothly covered by the probability density function.

In many theories, especially statistical physics and quantum theory, averages are granted a status close to that of the "real." From the perspective of generative ontology, however, averaging is not generation: averaging is only a statistical summary of many failures and a small number of successes. When a structure cannot be stably generated in a single process, its average value cannot provide it with ontological legitimacy. The real problem lies not in the use of probability, but in elevating probability into an ontological explanation. When a theory begins to claim that "the world is essentially random," "outcomes have no causes, only probabilities," or "generation is not the issue; distribution is the issue," then generative failure is formally declared "meaningless." Failure no longer requires explanation, because it has already been defined as the natural state.

The ontologization of probability almost always works in coordination with the continuous limit. A continuous spectrum permits infinitesimal differences, and a probability distribution permits infinitesimal-probability events. Combined, they form a structural buffer zone: any structure that does not appear can be assigned to the "low-probability tail," and failure therefore no longer constitutes a counterexample to the theory, but is merely something "not yet sampled." This logic is especially clear in numerical simulation: unstable structures are averaged after repeated runs, divergent trajectories can still be assigned statistical distributions, and processes that cannot close are explained as fluctuations or noise. Probability thereby becomes the final repository of numerical failure. Yet it must be emphasized that the repeatability stressed by probability theory is statistical rather than generative: a phenomenon may be statistically stable while failing in every attempt at generation, and such a phenomenon does not constitute "existence" in generative ontology.

This book therefore does not deny the use of probability, but strictly restricts its layer: in generative ontology, probability can appear only at the readout layer, describing the output statistics produced when phase information is gated and truncated by bandwidth, and it cannot in turn explain the generative mechanism—probability is the result of information loss, not the starting point of the world (this will be fully developed in Chapters 3 and 18). We can now see that the continuous limit, numerical approximation, and probabilistic averaging together constitute a dangerous theoretical posture: unfailability. The continuous limit defers failure to infinity; numerical approximation interprets failure as incompletion; probabilistic averaging statisticalizes and normalizes failure. Within this structure, the theory is always correct; we simply have not "looked finely enough." Generative ontology rejects this posture and requires the ontological status of failure to be restored: failure is not technical noise, not a statistical tail, but an ontological event. If a theory cannot specify under what conditions generation occurs, under what conditions it fails, and whether failure has irreversibility, then it has not yet genuinely touched existence. The temptation of probabilistic language lies in its extraordinary permissiveness, but it is precisely this permissiveness that causes a theory to lose its boundaries. When failure has been deferred by continuity, buffered by numerical limits, and then statisticalized by probabilistic averaging, it does not disappear; it only reappears in more extreme locations.

1.8. How the Container Assumption Spatializes Causality and Finally Becomes Visible in Three Predicaments

The influence of container grammar is not limited to "existence"; it also shapes the understanding of causality, a point often overlooked. When spacetime is treated as a container, causality is likewise spatialized and temporalized: "cause precedes effect" is understood as ordering along the time coordinate, and "influence transmission" is understood as propagation through space. This understanding is extraordinarily successful in engineering practice and empirical contexts, so much so that it appears to be causality itself. Yet it contains an untested premise: causal relations must attach to a preexisting spatiotemporal order—there must first be temporal succession and spatial distance before there can be causal "priority" and "transmission."

Once this premise is accepted, any phenomenon that does not conform to spatiotemporal intuition becomes troublesome. Quantum correlations, nonlocal correlations, and global constraints are not easily placed within the picture of "propagation through space and ordering in time." Faced with them, container grammar has only two options: either treat them as anomalies that must be explained away, or introduce higher dimensions, more complex geometries, or hidden variables to "repair" the container so that these phenomena can once again be placed within spacetime. Either way, the burden of explanation is outsourced to structure itself—anomalies are accommodated by making the structure more complex—while the generative logic of causal relations among events remains untouched. This suggests the opposite possibility: perhaps causality does not attach to spatiotemporal order, but rather the reverse—spatiotemporal succession and distance are readouts that appear after causal relations, that is, generative dependencies, have frozen in some sense. This chapter does not develop that possibility here, because it requires tools established in later chapters. It only points out that spatializing causality is another consequence of container grammar, not an independent empirical discovery; it is the other side of the same premise that spatializes existence.

At this point, a natural response emerges: if the problem lies in "continuous rigid spacetime," can the circle be avoided by adopting a more general concept of spacetime? For example, by using discrete spacetime, causal sets, spin networks, or various approaches that weaken the requirement of continuity? The answer of this book is a clear no, and this negative answer is the central point of the present diagnosis: so long as "spacetime," in whatever form, is still assumed as the precondition for the existence of events, the circle is not broken. The container may be replaced, geometry may be generalized, and continuity may be replaced by discreteness, but the direction in which "structure comes first and events come afterward" remains unchanged. A discrete causal set still begins by giving a partially ordered set composed of events and then discussing what occurs upon it; a spin network still begins by giving a network and only then discusses states upon it. These approaches make the container more refined and closer to discreteness, but do not abolish the ontological direction in which "the container precedes the content." The problem is not whether the container is continuous or discrete, but whether it is a premise.

This is why this book reads singularities as clues rather than defects—and singularities are only one of three locations where the same logical defect becomes concentratedly visible. In general relativity, the centers of black holes and the initial moment of cosmology correspond to so-called singularities: curvature diverges, the metric loses definition, and the standard statement is that "the theory fails here and requires quantum-gravity correction." But if spacetime is not a prior container, but a readout appearance produced after the successful freezing of a large number of relations, then a singularity is not a catastrophe in which the "container breaks," but an expected result: it marks the point at which the successful freezing that supports the spacetime readout no longer holds, and the underlying discrete generation reappears. The same structure also appears in two other places: the initial-condition problem outsources the whole question of "why generation occurred in this way" to an unexplained "starting point," effectively pushing responsibility for generation beyond the endpoint of the time axis; and quantum measurement collapse is the moment at which the container can no longer carry, on a given stage, the question of "how facts come to hold." All three are manifestations of the same thing: when a theory cannot complete generation within a given background but must nevertheless produce a result, rupture is unavoidable. Container grammar cannot tolerate any place where the container fails to hold, so it treats these three cases as embarrassments; once the container is abandoned as ontology, however, their existence becomes positive evidence—they are precisely the locations where the continuous readout withdraws and the generation layer becomes visible. The same phenomenon has entirely opposite status under two different starting points; this itself shows that the choice of starting point is not neutral.

Therefore, the conclusion of this chapter is not that "existing theories of spacetime are wrong," but a more restrained and more fundamental judgment: with respect to the problem of generation, any theory that takes spacetime as its starting point has already answered in advance the very question it should explain. Such theories have not answered incorrectly; rather, they have canceled the question—by writing the answer, namely the existence of spacetime, into the premise, they make the question of why spacetime exists and how events are generated impossible to pose within their own framework.

1.9. The Strategy of This Book: Suspension, Not Refutation

Precisely because the problem lies at the starting point of a theory rather than inside it, this book does not attempt to refute any specific model within existing frameworks. It adopts a stricter and also more conservative strategy: suspending spacetime as an ontological premise.

The meaning of "suspension" must be understood accurately, lest it be misread as a merely empty negation. Suspension does not deny the effectiveness of spacetime descriptions—this book fully acknowledges that, under appropriate scales and conditions, spacetime descriptions are precise and successful. Suspension means temporarily refusing to use spacetime to define events, causality, and existence. It removes spacetime from the position of "starting point" and places it in the position of a "result to be explained." Only by posing anew the questions "what is an event," "what does it mean to occur," and "how can something become stable," without relying on the container assumption, can generative logic be genuinely tested. Otherwise, the answers to these questions have already been covertly supplied by the container, and no test is possible.

This suspension is negative; it does not itself establish anything. Its sole function is to clear a position that has long been occupied by the container. In this position, an event is no longer presupposed as a point in spacetime, but can be reintroduced as a unit of generation whose occurrence remains to be adjudicated; causality is no longer presupposed as spatiotemporal order, but can be reintroduced as generative dependence; existence is no longer presupposed as "occupying a position," but can be reintroduced as a completed state that may either succeed or fail. These "reintroductions" constitute the positive work beginning in Part II; Part I is responsible only for clearing the position.

This position can now be considered genuinely cleared because this chapter has shown that container grammar does not occupy the starting point in isolation, but is supported layer by layer by continuity (1.3), axiomatization (1.5), numerical limits (1.6), and probabilistic averaging (1.7). Together, they form an "unfailable" theoretical posture that allows the "true starting point" to be deferred indefinitely. To clear the starting point, suspending spacetime alone is therefore insufficient; the entire structure that supports it must also be examined. This chapter has completed the diagnosis of that supporting structure; the deepest link in it, and the one most in need of separate treatment—the unfalsifiability of continuity itself—will be the task of the next chapter.

1.10. Summary of This Chapter

The aim of this chapter has not been to assess the success or failure of any specific physical theory, but to provide a structural diagnosis of a premise in modern physics that has long escaped systematic audit: events are assumed by default to occur within some preexisting structural background. This premise has never been proposed as an explicit proposition; instead, it has been consolidated layer by layer into an ontological commitment through descriptive language, formal modeling, and numerical practice.

The diagnostic chain unfolded as follows. At the most basic level of expression (1.1), once "events occur in spacetime" is adopted as a starting point, descriptive language is quietly substituted for an ontological premise. Once spacetime is established as a container, the formalization of a theory immediately enters a closed circle (1.2): spacetime is both the premise of events and the result of event evolution, self-consistent yet without a starting point. The assumption of continuity (1.3) provides the stabilizing mechanism for this circle—through infinite subdivision, limiting processes, and differentiable structures, it continually defers the problem of the starting point to ever smaller scales, thereby formally preventing the problem of generation from appearing explicitly. More broadly, this structure is not an accidental feature of any single theory, but the deep container grammar jointly inherited by Newtonian mechanics, relativity, and quantum theory (1.4): whether the background is called absolute space, a spacetime manifold, or a state space, its logical status remains the same—the background is assumed to exist, and the container has never been generated. Axiomatization (1.5) institutionalizes this premise at the formal level: existence is written into the axioms, generation no longer appears as a problem, and failure can only be understood as a deviation in derivation or computation, not at the level of whether an object has been generated. Numerical limits (1.6) further buffer the manifestation of failure in practice: what has not been generated is rewritten as "not yet converged," and responsibility for generation is deferred to unreachable limiting conditions. Probabilistic averaging (1.7) completes the final step of this rewriting: generative failure is statisticalized as the tail of a distribution and no longer appears as an object requiring explanation, giving the theory a structural immunity to failure. Finally, all these mechanisms become concentratedly visible in the spatialization of causality and in the three well-known predicaments of singularities, initial conditions, and measurement collapse (1.8). These are not independent technical difficulties, but manifestations of the same logical defect in different locations: when a theory cannot complete generation within a given background but must nevertheless provide a result, rupture becomes unavoidable.

The conclusion of this chapter is therefore not that "existing physics is wrong," but a more fundamental judgment: as long as events are still assumed to occur within some presupposed background, generation cannot become the responsibility of the theory itself. Continuous limits, axiomatization, numerical approximation, and probabilistic averaging jointly support an "unfailable" theoretical posture, and generative ontology rejects that posture—it requires failure to be restored as an ontological event. Chapter 1 has thereby completed its diagnostic task. The next task is no longer to repair structures within the container, but to suspend container grammar itself as a whole and begin again from generation, redefining the conditions of existence, structure, and failure. The deepest link in this task—why continuity itself is unfalsifiable—is precisely what the next chapter will first develop.

2. Continuity, Infinitesimals, and Unfalsifiability

The previous chapter suspended spacetime as an ontological premise and pointed out that this suspension is possible because what supports container grammar is not spacetime itself, but a deeper assumption: continuity. This chapter turns to that assumption. It does not seek to show that continuous mathematics is technically ineffective—on the contrary, it is extraordinarily successful in computation and engineering—but rather that continuity, infinitesimals, differential operators, and limits and renormalization jointly constitute, at the grammatical level, an explanatory structure that cannot fail, and therefore lose ontological efficacy with respect to the problem of generation. A framework that never has to acknowledge failure in the present cannot answer whether a structure has been generated; and this is precisely the final and most stubborn default that this book must first clear away.

2.1. Why the Continuous Limit Can Always Avoid Failure

Modern physics almost without exception begins with continuity: continuous time, continuous space, continuous fields, and continuous state spaces. Even in attempts at lattice formulation or discretization, the continuous limit is still preserved as the ultimate goal that should, in principle, be recoverable. This choice is usually explained as a matter of technical convenience, mathematical naturalness, or empirical success. This section argues something different: the central function of the continuous limit in theory is not to improve descriptive precision, but to systematically dissolve failure.

The decisive feature of continuity lies not in smoothness, but in a logical assertion: for any state, there always exists a closer state. Under continuous parametrization, this assertion appears as follows: after any moment, there is always a smaller time interval; near any position, there is always a closer position; any error can always be further reduced. Its direct implication is that there is no position at which one must, in principle, stop. And a framework with no place where it must stop also has no place where it must fail.

This property unfolds layer by layer into a buffering mechanism against failure. At the first level, divergence itself does not constitute failure: a field value tending to infinity is described as "divergence at the limit," increasing curvature is explained as "requiring a higher-energy-scale description," and the failure of an equation is attributed to "loss of applicability at that point." These formulations share the same structure: they defer failure to an unreachable limiting position, and as long as the limit has not actually been reached, the theory is still regarded as valid. At the second level, infinitesimals provide an unlimited buffer for every anomaly: any mismatch can be rewritten as "neglecting a smaller scale," "not including higher-order corrections," or "not yet reaching the true continuous limit." Failure therefore no longer means that a mechanism has not held; it only means that the approximation has not yet come close enough.

This mechanism directly conflicts with the requirement of falsification. Falsification requires a definite condition: under some attainable circumstance, the theory must fail. The framework of continuous limits systematically undermines this requirement—failed predictions can be met by further reducing the step size, numerical instability by refining the grid, and observational deviations by appealing to finite resolution. Continuity ensures that there is always an idealized limit not yet reached, and therefore always an escape route. The most common defensive phrasing of continuous theories is "in principle, under infinite precision..." or "in the limiting case, this behavior will appear..."; but "in principle" is not a physical condition, but a logical retreat. When a theory allows itself to hold only in an unrealizable limit, it detaches itself from the possibility of failing in reality.

In this way, the responsibility for generation is quietly replaced by the responsibility for approximation. From the perspective of generative ontology, a theory must answer: under what conditions is a structure generated, and under what conditions does generation fail? The implicit answer of the continuity model is: the structure always exists; it simply has not yet been resolved. This substitution produces a special explanatory temporal logic—the theory is always "close to correct," the computation is always "not yet converged," and the experiment always has "insufficient resolution." This logic has no endpoint, and therefore no moment at which failure must be acknowledged.

It must be emphasized that discretization does not automatically break this structure. As long as discretization is understood as an approximation to a continuous theory, the continuous limit remains the final arbiter, and the discrete structure does not acquire any genuine right to fail—any mismatch can be attributed to the discretization not yet being fine enough. The key issue is therefore not a technical dispute between the continuous and the discrete, but a grammatical choice: whether nature is continuous is an empirical question; taking continuity as the starting point of theory is a grammatical choice, and this choice is not neutral—it excludes, at the starting point, generative failure, structural rupture, and non-extendable termination. Generative ontology adopts the opposite starting point: evolution is not presumed extendable by default, closure may fail, and a process may terminate in an irreparable way. For this reason, the continuous limit cannot serve as the foundation of a theory of generation, but at most as an effective approximation of stable structures at the readout layer. Continuity is so successful in physics precisely because it is extremely permissive; the cost of this permissiveness is that theory loses the capacity to be forced into failure.

2.2. How Infinitesimals Dissolve the Boundary of Events

If the continuous limit provides a temporal and precision logic of "always being able to approximate one step further," infinitesimals provide the corresponding grammar of space and state: every change is allowed to be decomposed into countless minute differences that do not yet constitute events. This section shows that infinitesimals are not a neutral description of small-scale processes, but a modeling choice that actively dissolves eventhood.

In generative ontology, an "event" must satisfy at least two minimal conditions. The first is completion: once an event occurs, its internal state cannot be subdivided arbitrarily without changing its identity. The second is a decidable boundary: there must be a definite division between "before" and "after" such that "whether it has occurred" becomes a decidable proposition. Once these two conditions are undermined, the so-called event is no longer a unit of generation, but merely a marked position within continuous change.

The introduction of infinitesimals undermines precisely these two conditions, because it implies a strong logical commitment: no change is ever a completed occurrence, but is composed of infinitely many incomplete steps. In the language of differentials, change in position is written as $\mathrm{d}x$, temporal evolution as $\mathrm{d}t$, and change of state as $\mathrm{d}\psi$; what these quantities share is that they do not themselves constitute events, but only indicate potential directions of change. The boundary of an event is thereby systematically rewritten as a slope: "whether a transition has occurred" becomes "whether the rate of change is large enough," "whether a structure has formed" becomes "whether the gradient is steep enough," and "whether generation has been completed" becomes "whether the derivative is continuous." The judgment of the existence of an event is thus replaced by the smoothness condition of a function.

Within this grammar, generative failure is logically excluded. In generative ontology, failure means that, after a finite step, a structure can no longer continue to generate; but within the infinitesimal framework, a stalled evolution is expressed as the derivative tending to zero, structural rupture as a function becoming nondifferentiable, and model failure as "non-applicability at that point." In all these cases, failure does not terminate the process, but merely changes the mode of mathematical description. It is especially important to point out that nondifferentiability is not the same as generative failure: in differential language, nondifferentiability only means that one must change coordinates, introduce higher-order corrections, or adopt piecewise or weak solutions. It is a technical signal, not a generative termination. Infinitesimal grammar therefore naturally supports a logic of "always being repairable": renormalization near singularities, distributions at discontinuities, and weak solutions at abrupt changes. These treatments are all technically legitimate, but they share the same premise: no rupture is allowed to count as a terminal condition of generation.

As a result, the event is reduced to a reference point: a parameter value on a curve, a point on a manifold, or a position in a solution space. Such an "event" has no intrinsic completion; its identity depends entirely on an external coordinate system, and it is not ontologically equivalent to an event understood in generative ontology as a completed unit. Infinitesimals also form a structural alliance with the language of measurement—any measurement error is attributed to finite resolution, and any deviation is explained as unresolved minute variation. Measurement can therefore only approximate an ideal value without limit, but cannot identify a node of fact that has already been completed. Ultimately, "termination," a concept central to generative ontology, is systematically dissolved: time is infinitely subdivided, states are infinitely interpolated, paths are infinitely extended, and termination is rewritten as "tending toward a limit" rather than "stopping here."

This book therefore adopts a definite position: infinitesimals can serve only as approximation tools at the readout layer; they cannot serve as the ontological language of the generation layer, still less can they define whether an event exists. Generative ontology requires events to possess discrete completion, closure to be decidable, and failure to be irreparable—whereas infinitesimals grammatically prevent events from becoming fundamental objects. They rewrite the world as boundaryless change, interminable evolution, and an approximating process incapable of bearing failure.

2.3. How Differential Operators Rewrite the Problem of Generation as a Problem of Evolution

Continuity provides the temporal background of "in principle, indefinite extension"; infinitesimals provide the unit of change that is "always incomplete"; and differential operators complete the final step of this grammatical chain: they systematically rewrite the question "whether a structure is generated" as the question "how a state changes with a parameter." Once differential operators become the basic language of a theory, physical theory no longer has the capacity to ask "whether generation occurs," but can only ask "how evolution proceeds."

These two questions belong to different ontological categories. The problem of generation is adjudicative: will a certain structure appear? Does its appearance depend on closure conditions? Under what conditions does generation fail? The answers to these questions can only be "holds/does not hold/fails." The problem of evolution asks something else: given an initial state, how does the system change? How does the state move continuously with a parameter? Does the solution exist, is it unique, and is it stable? The issue here is not "whether something occurs," but "how something changes." At the grammatical level, differential operators imply three strong premises: the state already exists, since differentiation can act only on already defined functions, fields, or states; change can be locally linearized, since any change can be characterized by a local slope; and evolution can unfold in time, since change can be written as a stepwise advancing process. Taken together, these three premises mean that the problem of generation has already been grammatically excluded before it can be posed.

Consider a typical differential equation, $\mathrm{d}x/\mathrm{d}t = F(x,t)$. This expression already presupposes that $x(t)$ exists, that $x(t)$ is differentiable, and that evolution is controlled by a continuous time parameter. The only permissible question left to it is therefore "what is the trajectory of the solution," not "why does this trajectory exist" or "whether it might not exist." All questions related to generation are systematically transferred to "initial conditions": why did the structure appear? Because the initial conditions were such; why this state? Because the initial value was so specified; why not another? Because the initial conditions were different. Initial conditions thereby assume the entire function of closure criteria in generative ontology, without having to provide any generative mechanism.

This also makes evolution equations structurally unable to fail. In differential theory, "failure" can only appear as nonexistence of solutions, nonuniqueness of solutions, or divergence at some point; but all of these are treated as mathematical problems rather than physical failures—if a solution does not exist, change the function space; if it is nonunique, add regularity conditions; if it diverges, introduce renormalization or truncation. In other words, the evolutionary process can always continue, only in a different form. Differential operators structurally guarantee a kind of perpetuity: the time parameter can always be extended forward, the state can always be updated, and the equation can always be made piecewise or modified. Even at so-called "endpoints"—singularities, phase transitions, or moments of measurement—differential language automatically rewrites them as intervals of non-applicability, parameter jumps, or piecewise evolution.

Within this language, generation is usually renamed phase transition, bifurcation, or change of stability. Yet these concepts still belong to the category of evolution: phase transition is the result of parameter change, bifurcation is a change in solution structure, and stability is a dynamical property. They describe how already generated structures change, not why a structure is generated here. Differential time is a parameter axis that can be infinitely subdivided; it contains no natural moment of completion, no irreversible point of settlement, and no reason why it must stop. History is therefore expressed as a continuous trajectory that can be extended and interpolated at will. What generative ontology requires is precisely what is logically incompatible with this: nondifferentiable points of completion, closure events that cannot continue, and irreparable cases of failure.

The position of this book can therefore be stated clearly: evolution can describe changes in already generated structures, but it cannot explain why structures are generated, still less can it adjudicate whether generation fails. Differential operators are not erroneous tools, but at the grammatical level they systematically avoid the problem of generation—they express the world as the continuous change of what already exists, an unending process, and a history that can always be repaired. For this reason, at the generation layer, differential operators must be suspended, continuous time must be replaced, and the language of evolution must yield to the language of closure.

2.4. Limits and Renormalization: Deferring Failure as "Not Yet Complete"

After continuity, infinitesimals, and differential language have been established, modern physics still requires a mechanism to handle its internal tensions—divergence, singularities, indefiniteness, and infinities. That mechanism is limits and renormalization. The central claim of this section is that limits and renormalization do not repair failure, but defer it; they systematically rewrite theories that should fail under finite conditions as theories that are "not yet complete."

In mathematics, limits have a clear semantic function: they do not require an object to exist at a point, but only require that it can be approached indefinitely in a neighborhood of that point. Therefore, when a quantity diverges at a point, the language of limits permits us to say that the quantity does not fail to exist, but exists "in the limiting sense." In this semantic conversion, a crucial step has already been completed: failure is rewritten as non-arrival. Questions that are adjudicative in generative ontology—whether closure occurs, whether completion occurs, whether failure occurs—are thereby uniformly rewritten as whether something approaches, whether it holds in the ε–δ sense, or whether it can be extended in some topology. The judgment of "yes/no" is replaced by "arbitrarily close without requiring arrival." Singularities are the most typical example of this transposition: they should originally be treated as signals of theoretical failure—the equation is no longer applicable, physical quantities lose definition, and prediction cannot continue—yet in the language of limits they are redescribed as coordinate pathologies, parametric limits, or regions requiring a higher-energy-scale description. The singularity is no longer a terminal point, but is transformed into an interface leading to the next layer of theory.

Renormalization further structures this logic. It is not a single technique, but a set of stable commitments: divergence is not an error, but a scale that has not yet been separated; infinity is not failure, but a parameter that has not yet been redefined; noncomputability is not termination, but a need for recalibration. Its core conviction can be summarized in one sentence: as long as enough scales and redefinitions are introduced, self consistency can always be restored. Its most far-reaching effect is to introduce a new criterion of legitimacy—physical acceptability—and thereby produce a stable split: mathematically divergent, yet renormalizable; failing strict definition, yet still regarded as yielding valid physical predictions; formally unclosed, yet numerically matching experiment. Within this structure, failure is systematically moved outside the interior of the theory.

It is in the joint grammar of limits and renormalization that a stable physical intuition gradually forms—"always just short of completion." The theory is incomplete because a sufficiently high energy scale has not yet been reached; the problem is unresolved because the limit has not yet been fully taken; the deviation exists because the order of renormalization is insufficient. The decisive feature of this intuition is that it always permits a next step; and as long as the "next step" is always permitted, failure never has to be acknowledged in the present. The framework of "effective theory" institutionalizes this logic of deferral: a theory is valid at a certain scale, and beyond that scale it bears no explanatory responsibility; failure is uniformly transferred to an unknown high-energy regime. This practice is legitimate in application and indeed extraordinarily successful; but its conceptual cost is clear—the question of whether generation has occurred is systematically suspended rather than answered. Unfalsifiability here is not an accidental flaw of some theory, but the inevitable product of this grammar: every counterexample can be interpreted as a limit not yet reached, every deviation absorbed into a parameter redefinition, and every failure deferred to a higher scale.

By contrast, generative ontology insists on several non-negotiable minimal requirements: closure either holds or does not hold, completion is not an asymptotic process but an event, and failure must be decidable and irreparable. Therefore, any structure that relies on continuous limits, infinite redefinitions, or perpetual extension must be suspended at the generation layer. The position of this book is clear and uncompromising: a theory that does not allow real failure cannot explain generation; genuine generation must be accompanied by non-extendable breakpoints, non-renormalizable collapse, and cases of failure that cannot be avoided through parameter adjustment.

2.5. Why "Existence of a Limit" Is Not a Physical Proposition

The preceding section considered how limits and renormalization handle failure in practice. This section asks a more basic question: does such handling constitute a physical explanation semantically, or does it merely perform a linguistic rewriting from "failure" to "mathematical existence"? This distinction is decisive for the problem of generation. The claim of this section is that "existence of a limit" is not a physical proposition, but a statement of legitimacy internal to mathematical language. Confusing the two is one of the key reasons why modern physics repeatedly falls into a circle with respect to the problem of generation.

In mathematics, "existence" has a clear and limited meaning: existence within an axiomatic system, satisfying given definitions and logical conditions, without requiring realization, occurrence, or completion. When we say that $\lim_{x\to 0} f(x)$ exists, its meaning is only that for every ε, there exists a δ such that the corresponding inequality holds—this is a purely logical-semantic proposition. A physical proposition must satisfy at least one minimal condition: it must point to some process that has occurred or could occur. A physical proposition cannot merely describe "what would happen if one approached without limit"; it must answer whether completion occurs, whether settlement occurs, and whether a retainable fact is left behind. "Existence of a limit" answers none of these questions.

The language of limits has a structural feature: it suspends time and occurrence. A limit is not a moment, not an event, and not an identifiable node of process; it describes an endpoint that need not be reached. Thus, expressions common in physical texts such as "in the limiting sense, this quantity is finite/definable/well behaved" are perfectly legitimate mathematically, but semantically incomplete physically. They do not state whether there is a physical process that completes the limit, whether there is a mechanism for determining that the limit has been realized, or what happens if it cannot be realized. When a limit is mistaken for a physical statement, it quietly replaces the role of the "event": closure is no longer required, completion is no longer required, freezing is no longer required; the ability to "approach without limit" seems sufficient. At a deeper level, what a limit often describes is a world that has never occurred—no system truly undergoes infinitely many steps, no process completes infinite approximation, and no event settles at a "limit point." A limit is therefore the language of an observer, not the generative language internal to the system.

This also determines why limits cannot bear falsification or causality. For a proposition to be falsifiable, it must at least have failure conditions, failure must not be eliminable by linguistic extension, and failure must have a definite criterion of adjudication. "Existence of a limit" does precisely the opposite—failure can be explained as "not close enough yet," deviation as "the limit has not yet been fully taken," and counterexamples are systematically deferred; it is therefore structurally incapable of failure. Limits likewise provide no causal explanation: they do not explain why something approaches, what drives the approach, or how the process terminates if the approach fails. They are only semantic packaging of the result, not a generative mechanism. Generative ontology therefore proposes a minimal alternative requirement: if a limit is to be used, one must specify its generative path of realization, identify its closure criterion, allow it to fail, and state the consequences of failure. Otherwise, the limit can only be treated as a mathematical tool, not as a statement of physical ontology. On this basis, this book states clearly: any claim grounded solely in the "existence of a limit" does not constitute a physical generative proposition. In this book, physical existence means generative completion; holding means closure settlement; explanation means mechanism plus a path capable of failure. Any "existence" that fails to satisfy these conditions does not enter the ontological layer.

2.6. Mathematical Differentiability Is Not Physical Generability

This section does not introduce a new tool, but draws the preceding sections together at the level of criteria: in the sense of generative ontology, mathematical differentiability does not guarantee physical generability in any way. There is a default equation in modern physics that has almost never been directly questioned: "as long as an equation is differentiable, it describes a physically generable process." This equation is neither a mathematical theorem nor a physical conclusion, yet it is widely used as a premise.

The meaning of differentiability in mathematics is limited and clear: in a neighborhood of a point, a function can be approximated by a linear map, with the error controlled in the limiting sense. This property concerns only the local structure of a function, and does not concern whether a process occurs, whether a state is completed, or whether a system can reach that point. It describes representational capacity, not generative capacity. For a physical process to be called generable, it must satisfy at least one of the following: there is a realizable path of evolution, there is a criterion of completion or settlement, or there is the possibility of failure, bifurcation, or termination. "Differentiable at a point" guarantees none of these. It guarantees only that, if the point is reached, a linear expansion can be made in its neighborhood. Many mathematically well-behaved objects are not physically reachable—infinitesimal time steps, infinitely fine spatial resolution, arbitrarily precise state control. Differentiability allows them to exist in the language of description, but does not require the system to truly undergo them in the generative sense.

This exposes a deep confusion: treating "giving a differential equation" as equivalent to "giving a generative mechanism." In fact, a differential equation only specifies "if the system is already in this state, how does the rate of change behave"; it does not specify whether the system can enter that state, nor whether there exists a physical process that completes an "instant." Differential form presupposes an already existing and infinitely divisible evolutionary parameter, so the problem of generation is quietly placed in advance and concealed. When a theory is built upon a differentiable manifold, it implies at least three premises: structure continues to hold at arbitrarily small scales, state space remains isomorphic across all scales, and the evolutionary parameter can be infinitely subdivided without generative cost. These are mathematical conveniences, not physical conclusions of generation. The same distinction must be made rigorously in the theory of partial differential equations: "existence of solutions" is a central problem, but the existence of a solution is not the existence of a process, uniqueness of a solution is not uniqueness of evolution, and regularity of a solution is not generative realizability. "A solution exists" means only that a certain function satisfies the equation; it does not mean that the universe will generate that function.

Differentiable structures have another important side effect: they systematically dissolve failure. If a point is unreachable, one can go around it in a neighborhood; if a behavior diverges, it can be treated through rescaling; if a structure is unstable, it can be linearized in a smaller neighborhood. Failure no longer appears as a rupture of generation, but is rewritten as "local bad behavior." Generative ontology makes a clear assertion here: any structure that cannot be discretized, closed, or frozen does not constitute a physical object of generation even if it is mathematically differentiable. Differentiability can serve only as a local approximation of steady structures, an effective tool at the readout layer, and a descriptive language after generation has been completed; it cannot serve as an ontological starting point. This also requires a systematic reversal of a common value judgment. Traditionally, smoother solutions are often considered better theories, and fewer singularities more successful models; in generative ontology, however, smoothness does not guarantee generative legitimacy, singularities often expose generative rupture, and nondifferentiable locations may correspond to real nodes of settlement. "Smoother" does not mean "more physical."

2.7. Summary of This Chapter: Why the Continuous-Differential Language Cannot Bear Responsibility for Generation

This chapter has systematically analyzed the joint operation of continuity, infinitesimals, differential operators, and limits and renormalization in modern physics. Its conclusion is not that these tools are technically ineffective, but that they jointly constitute, at the grammatical level, an explanatory structure that cannot fail, and therefore lose ontological efficacy with respect to the problem of generation.

The operation of this structure proceeds layer by layer. Continuity introduces, through infinitely divisible parametrization, a temporal logic of "always being able to approximate one step further," so that evolution no longer has nodes at which it must terminate or settle. Infinitesimals further dissolve the boundaries of events, rewriting generative processes that should possess completion as the sum of countless incomplete changes; events thereby degenerate into reference points on trajectories rather than real units of generation. On this basis, differential operators perform the key rewriting, replacing the question "whether a structure is generated" with the question "how a state evolves with a parameter": generation is outsourced to initial conditions, failure is rewritten as a mathematical case, evolution is granted the legitimacy of perpetual extension, and physical theory thereby loses the grammatical position from which to ask "whether generation occurs" and "when it fails." Limits and renormalization do not repair this gap, but institutionalize it—through narratives such as "holding in the limiting sense" and "being valid at a higher energy scale," generative failure is deferred as incompletion, and theoretical responsibility is transferred to unreachable ideal limits. The framework of "effective theory" further consolidates this deferral, keeping failure permanently outside the theoretical field of vision. Within this entire structure, judgments such as "existence of a limit," "differentiability of a solution," and "mathematical well-behavedness" are frequently misread as grounds for physical existence. This chapter has clearly distinguished them: mathematical existence is not equivalent to physical occurrence, and differentiability does not guarantee generability. Such judgments belong at most to the readout layer or the descriptive layer, and cannot serve as ontological premises of the generation layer.

The conclusion of this chapter is therefore clear: any framework that takes continuity, infinitesimals, differential evolution, or the existence of limits as its starting point grammatically excludes the possibility of real failure, and therefore cannot bear the task of generative explanation. This does not deny the effectiveness of continuous physics, but strictly limits its layer of application—the continuous-differential language can only describe how already generated and frozen structures change; it cannot explain how structures are generated, why they close, or under what conditions they fail. In the next chapter, this book will no longer attempt to repair this language, but will suspend it completely and turn to the third and final unexamined default of modern physics: probability. Only by clearing the container, continuity, and probability one by one can the problem of generation first become an object that is decidable, capable of failure, and genuinely testable.

3. The Suspended Status of Probability

In modern physics, few concepts are used so frequently, accepted so widely, and yet never given a clear definition at the physical level as "probability." It appears in statistical mechanics, quantum mechanics, field theory, cosmology, and information theory; it is calculated, manipulated, fitted, and cited, yet it always exists in a peculiar way: it never appears as a physical object that can be generated and can fail. This chapter is the final step in the diagnosis of Part I. It aims to show that probability remains suspended not because it is useless, but because it has been placed at the wrong layer—a statistical tool that properly belongs to the readout layer has been misplaced as an ontological component of the generation layer, and therefore can neither be defined, nor fail, nor be generated.

3.1. Probability Has Never Been Defined in Physics

In physical theory, fundamental quantities usually have definite ontological status: position, momentum, energy, frequency, and phase all correspond to definable states, processes, or readout interfaces. Probability alone is an exception. It has never been specified as a field, an evolving process, a state variable of a system, or a structure that can freeze or fail; it belongs neither to the system itself nor to the physical relations among systems. This is an anomalous fact that deserves direct attention: probability is everywhere, yet it has no ontological position.

This is partly because probability has been interpretively unstable from the beginning. In physical practice, it is usually assumed to mean one of the following: the propensity of an event to occur, the frequency limit of infinitely repeated experiments, the incompleteness of an observer's knowledge, or a weight in a mathematical measure. The problem is that these four interpretations are not equivalent, and at least three of them are not physical concepts. Interpreting probability as a frequency limit appears empirically robust, but it implies three unstated premises: the experiment can be repeated infinitely, each repetition is physically "the same," and the limiting frequency genuinely exists. From the perspective of generation, however, every act of generation is an independent event of settlement; "repetition" itself must be generated rather than assumed, and frequency describes regularity in posterior statistics rather than the mechanism by which an event occurs. Repeatability is not generability. Interpreting probability as incompleteness of knowledge locates it in epistemology: this is reasonable at the readout layer, but it cannot support the ontologized use of probability—when a theory claims that "the world is essentially probabilistic," it has already substituted a statement about knowledge for a statement about existence. Epistemological uncertainty cannot prove ontological randomness. Interpreting probability as a weight on a measure space is the mathematically cleanest treatment, but also the one that most thoroughly avoids the physical question: a measure only specifies how weights are assigned to an already given set of events; it explains neither why these events appear nor why the weights take this form. Mathematical weight is not physical occurrence.

A more structural fact is that, in the fundamental dynamical equations of all mainstream theories, probability itself never appears as the quantity being evolved. The Schrödinger equation evolves the wave function rather than probability, and Hamiltonian dynamics evolves the support of a phase-space distribution rather than probability itself; probability is always appended outside the dynamics through an additional rule, such as the Born rule. This shows that probability is not a component of the generation layer, but an externally added rule of readout. According to the methodological principle of this book, any fundamental physical quantity must be able to fail—there must exist some decidable condition under which it does not hold. Yet probability is never allowed to fail: every deviation can be attributed to fluctuation, and every counterexample deferred to a larger sample. A "fundamental quantity" that can never fail has no ontological qualification in generative terms.

Probability therefore displays the typical symptoms of a "suspended concept": it is universally used but has no unified definition, is treated as ontological but never enters dynamics, is regarded as fundamental but cannot fail, and is simultaneously supported by multiple mutually nonequivalent interpretations. These symptoms jointly point to a single judgment: probability has not yet acquired an ontological position in physics. The position of this book is that probability indeed has a legitimate place, but not at the generation layer; it belongs to the readout layer. It is not a property of the world, but a consequence of finite readout operations. The full argument for this position will be developed in Part IV, especially Chapter 18. The task of the present chapter is only to clear in advance the illegitimate position of probability: when a statistical tool of the readout layer is elevated into an ontological component of the generation layer, it can neither be defined, nor fail, nor be generated. Clarifying this misplacement is the work of the remaining sections of this chapter.

3.2. Why Probability Appears Only After Measurement

There is a simple yet decisive empirical fact: in any actual physical situation, a probability distribution is defined only after measurement. Before measurement occurs, the system is described by a determinate state, such as a wave function or phase structure; only when measurement intervenes do results appear in probabilistic form. This temporal order is not accidental; it is itself a clue to the layer to which probability belongs.

It is worth noting that "measurement" is not described by the fundamental dynamical equations in standard theory. The evolution equation is deterministic and reversible, whereas measurement is formulated as an additional, non-unitary, and irreversible step. This means that probability appears precisely where dynamics is interrupted and another kind of operation intervenes. Such operations share a common physical feature: they are constrained by three kinds of finitude—a finite time window, finite phase resolution, and finite spectral bandwidth. These three kinds of finitude jointly constitute a truncation of the generative structure, and probability appears precisely at this point of truncation. (The exact mechanism of this triple truncation will be given systematically in Part IV as the core structure of the readout layer.) Under truncation, complete phase information in the generation layer cannot be fully preserved. The discarded part does not disappear, but appears in the form of statistical dispersion. A probability distribution is precisely the statistical residue left after phase information has been truncated—it measures how much has been discarded, not how random the world itself is.

If probability is the product of truncation, then it cannot logically exist before truncation. Before generation has been truncated by any readout protocol, there is no "probability" to speak of, only determinate phase evolution. This explains why probability always appears after measurement: before measurement, truncation has not yet occurred. The role of measurement is thereby reinterpreted—it does not reveal a random result that already existed, but is a gating operation. It projects a generative structure through finite bandwidth and a finite window, discarding phase details in the process; probability is the by-product of this gating, not a preexisting fact revealed by it.

Two empirical indications reinforce this judgment. First, the same system yields different probability distributions under different measurement bases and different measurement settings. If probability were an intrinsic property of the world, this variability depending on the mode of measurement would be difficult to explain; if probability is the product of readout truncation, its variation with the readout protocol is precisely what should be expected. Second, and more strongly, probability distributions can be systematically modulated by changing measurement conditions—altering the gating time, phase resolution, or bandwidth deforms the distribution accordingly. This modulability is key evidence that probability belongs to the readout layer: ontological properties of the world should not change with the way we read them, but the products of readout should. Taken together, these points yield a clear positional judgment: probability appears after generation has been completed and when a readout protocol intervenes; it does not appear at the generation layer, because the generation layer contains only determinate phase evolution and closure adjudication. Appearing only after measurement, varying with the mode of measurement, and being modulable by measurement conditions—these three indications jointly point to the same conclusion: probability belongs to the readout layer, not the generation layer.

3.3. Why "Probabilistic Explanation" Cannot Be Falsified

For an explanation to be falsifiable, it must satisfy at least one minimal criterion: there must exist some attainable circumstance under which the explanation must fail, and this failure must not be removable through linguistic restatement. This section shows that ontologized probabilistic explanations precisely fail to satisfy this criterion—and this is not their advantage, but a signal that they have been placed at the wrong layer.

When probability is placed at the ontological layer, it is equipped with a standard escape grammar: deviations are explained as "statistical fluctuations," counterexamples as "low-probability events," and mismatches as "insufficient samples." This grammar ensures that any empirical conflict can be absorbed. One might expect that enough data would eventually falsify an incorrect probabilistic claim; but the escape grammar of ontologized probability defeats this expectation—any finite dataset, no matter how large, can still be interpreted as not having reached the limiting frequency, and "more data" can always be demanded. Falsification can therefore always be deferred. The true function of ontologized probability is thus not prediction, but retrospective tolerance: it reserves explanatory space for almost any observational outcome, and a statement that can tolerate every outcome does not constitute a testable prediction.

The Born rule $P = |\psi|^2$ is often regarded as the foundation of probability in quantum theory, but its logical status deserves scrutiny: it is a rule appended outside the dynamics, is not derived from the dynamics itself, and does not explain why probability takes this form. It specifies "how to calculate probability," but avoids the question "why probability appears." When asked why the world is probabilistic, ontologized explanations often appeal to the fact that "measurement gives probabilistic results," while the question "why does measurement give probabilistic results" is answered by saying "because the world is probabilistic." This is a circle: probability is used to explain its own appearance.

It must be emphasized that this book does not criticize unfalsifiability merely as a logical defect, but treats it as a warning signal: it shows that probability has been placed at the wrong layer. A correctly positioned concept of probability should be able to fail. This issue has long escaped direct treatment partly because probability is computationally extraordinarily successful—the numerical values it gives agree with experiment to a high degree; but computational success conceals ontological ambiguity: calculating correctly does not mean understanding. This book therefore proposes a clear requirement: if probability is to have physical meaning, it must be placed somewhere capable of bearing failure—there must exist decidable conditions under which the claim that "probability arises from this mechanism" can be overturned. That position is precisely the readout layer. The fundamental problem with ontologized probabilistic explanation is not that it is wrong, but that it is immune to failure: it can explain everything, and therefore it truly explains nothing. To restore its physical meaning, it must be downgraded to a layer capable of failure.

3.4. The Only Legitimate Position of Probability: A Truncation Effect at the Readout Layer

The preceding three sections have shown that probability can neither be defined at the generation layer, nor fail there, nor be generated there. This yields an exclusionary conclusion: if probability has a legitimate position at all, it can only appear at a lower, operational layer. This layer must satisfy three conditions at once: probability does not participate in generative rules; probability depends on a specific operational protocol; and probability can be systematically weakened, deformed, or eliminated when conditions change.

The scope of this section should be made clear: its task is not to develop the full mechanism of this layer, but to establish its position and deliver the positive argument to the later text. Its core claim can be stated clearly: probability is not a property of the world, but an inevitable by-product of the readout process.

In terms of layer assignment, any physical theory contains at least two domains of questions that must not be conflated: the domain of generation, concerning how structures appear, close, and freeze; and the domain of readout, concerning how generated results are recorded, compressed, and output by finite systems. Traditional theories structurally conflate the two—when generation cannot be explained, probability is written directly into the domain of generation and thereby replaces mechanism. This book explicitly rejects that substitution and strictly relocates probability to the domain of readout. Here "readout" has nothing to do with consciousness or subjectivity; it is defined as a phase–intensity gated projection operation with finite bandwidth, finite time window, and finite resolution. Its elements—gating time, phase fidelity, bandwidth, and threshold rules—are all physical parameters that can be adjusted, compared, and can fail.

This relocation yields the decisive gain of the present section: falsifiability. Once probability is downgraded to the readout layer, it acquires, for the first time, the capacity to bear failure. Its falsifiability can be stated preliminarily in directional form: increasing phase fidelity should shrink the probability distribution or reveal structure; shortening the gating time should cause the statistics to reveal phase structure; changing the bandwidth should produce systematic drift in the distribution. If none of these controlled changes appears, then the claim that "probability arises from readout truncation" fails. This is a testable and refutable claim, not an unfailing ontological stipulation. (The operational threshold form of this falsification criterion—involving measurable proxies such as coherence time, visibility, and the CHSH parameter, together with order-of-magnitude thresholds—is part of the same line of argument as the fourth item in Part Zero, "How to Falsify This Theory," and the Generation-Layer Accessibility Proposition. Its full statement is given in Chapter 18 and Appendix H.)

The allocation of argumentative position must be made explicit, so that readers do not expect this chapter to provide what does not belong to it. The full positive construction of this layer—the exact mechanism of triple truncation in the readout protocol, why under specific truncation $P(x) \propto |\psi(x)|^2$ appears as a minimal readout invariant, and the calculable drift law of probability under changes in the readout protocol—all belong to the systematic content of the readout layer and are reserved for Part IV. Why the readout layer must be introduced is discussed in Chapter 14; the readout origin of time, space, and causality is discussed in Chapter 15; and the full mechanism of probability as a truncation phenomenon, together with its falsification card, is discussed in Chapter 18 and Appendix H. This chapter does not repeat those derivations, in order to preserve the diagnostic purity of Part I: Part I explains why probability cannot serve as an ontological starting point; Part IV will explain how it arises as a readout residual through truncation of generation.

This positioning also clarifies how this book differs from traditional interpretations of quantum theory—the difference lies only in layer assignment. The Copenhagen interpretation treats probability as a property of nature, while the many-worlds interpretation treats probability as branch weight; both place probability at the ontological layer. This book presses probability back to its only legitimate position, namely the truncation effect of readout. The disagreement is not over empirical conclusions, but over layer assignment. This section has therefore not denied the effectiveness of probability, but has completed a strict layer reset: from ontology down to operation, from starting point down to result, from unfailing down to adjustable, testable, and refutable. Only in this position can probability become part of physics again; how it arises in this position through the truncation of generation is the task of Part IV.

3.5. Summary of This Chapter: Reassigning Probability to Its Proper Layer

The task of this chapter has not been to propose a new interpretation of probability, but to accomplish something more fundamental: to determine the legitimate position of probability within physical theory and to exclude its illegitimate position.

The step-by-step analysis yields a clear conclusion: probability is widely used in modern physics, yet it has never acquired a definition at the generation layer; it is neither an evolutionary variable, nor a closure criterion, nor does it bear any responsibility for generation. The chapter first identified this crucial fact (3.1)—probability is everywhere yet remains suspended, and neither frequency interpretations, nor knowledge interpretations, nor measure interpretations can answer why probability appears, when it appears, or what happens if it does not appear. The chapter then provided an empirical adjudication of where probability appears (3.2): probability appears only after measurement, varies with the mode of measurement, and can be modulated by measurement conditions. These three indications jointly point to the same conclusion—probability belongs to the readout layer, not the generation layer. Further analysis showed (3.3) that the reason probability is difficult to falsify is not that experiments are insufficient, but that once it is placed at the ontological layer, it structurally acquires immunity to failure: all deviations can be explained as fluctuations, and all counterexamples can be deferred to larger samples or higher precision, causing probabilistic explanation to withdraw from the falsification system of physical theory.

The core turn of this chapter lies in resetting the layer assignment (3.4): probability is not an attribute of generation, but a consequence of readout. It appears only after generation has been completed and when a readout protocol intervenes, and it depends strictly on operational conditions such as gating time, phase fidelity, bandwidth, and threshold. In this position, probability for the first time becomes adjustable, comparable, and capable of failure. It must be emphasized again that this chapter only positions this layer and does not develop it: the full generative mechanism of probability as a readout residual, its readout invariant, and its falsifiable degradation law belong to Part IV (Chapters 14, 15, 18, and Appendix H). They are not repeated here, in order to preserve the purity of Part I as diagnosis.

This chapter has therefore not denied the effectiveness of probability, but has performed a necessary downgrading: from ontological attribute to operational result, from starting postulate to truncation effect, and from unfailing to conditional product. This downgrading is decisive—it cuts off probability's concealment of generative failure and reopens the logical space for the questions of whether events are generated, whether structures close, and whether freezing occurs. At this point, Part I has completed its clearing of the three defaults of container, continuity, and probability. Beginning with the next chapter, the argument will no longer rely on the language of probability, but will enter directly into the ontological layer of generation, taking events, relations, closure, and freezing as its only starting points.

Part II.

Generative Ontology: The Event as the Only Starting Point

Introduction

The conclusion of Part I is not a declaration of position, but a clear operational result: any physical formulation that begins from continuous spacetime, a probability postulate, or an unfailing limit is insufficient to bear the role of a theory of generation. This judgment does not depend on whether any specific model succeeds, but follows from a systematic examination of its logical starting point—the container smuggles existence in as a premise, continuity defers failure as "not yet complete," and probability gains immunity to failure at the ontological layer. Once these three are suspended, what remains is not a new theory, but an emptied starting point. The task of Part II is to begin positive construction from this emptied starting point.

The nature of this construction must first be clarified, lest it be misread as another renovation of concepts. Part II is not a matter of "replacing variables" or "reinterpreting concepts" within an existing physical framework. Rather, it explicitly abandons three kinds of starting points long regarded as self-evident: spacetime as an a priori container that bears events; probability as a component of dynamics or ontology; and the continuous limit as a guarantee of generative legitimacy. After abandoning these three, the problem addressed in this part is strictly reduced to a minimal generative question: if spacetime is not presupposed, probability is not introduced, and continuity is not relied upon, then what structure is minimally required for "occurrence"?

This question requires a thorough retreat in the order of exposition. Traditional physical theories unfold around "laws": given a state space, one specifies evolution equations on it and then generates results through initial conditions. Part II deliberately steps back before this structure, not because the laws are necessarily wrong, but because laws presuppose differentiable variables, states presuppose comparable moments, and evolution presupposes continuous parameters—and these are precisely the contents that Part I has systematically suspended. Therefore, Part II does not discuss "laws," but the grammar of generation: what kinds of rules qualify as physical generative rules.

Within this grammar, the basic components are no longer particles, fields, or states, but events. The term "event" here has a strictly limited meaning. It is not a spacetime point, not a measurement result, and not a change in the properties of an object; an event is defined as an indivisible unit of generation–settlement, whose existence is adjudicated by closure conditions. The key point of this definition is that the existence of an event is not a logical premise, but the output of a generative process. In contrast to traditional dynamics, Part II does not begin from equations of evolution, but from an adjudicative logical order—whether generation occurs; whether generation can close within a finite number of steps; what can close is frozen as fact; what cannot close is eliminated, decays, or dissipates. In this order, closure is not a result of evolution, but the threshold of existence; freezing is not a final state, but the condition of reproducibility; failure is not an anomaly, but a normal output of the generative process.

It is precisely the concept of "failure" that marks the deepest divergence between this part and traditional theory. In traditional theories, failure usually means model error, numerical instability, or insufficient approximation—it is a defect waiting to be repaired. In this book, failure is elevated into a fundamental generative criterion rather than a rhetorical term. Part II will systematically introduce failure as a structural element and require every generative rule to satisfy three conditions: it can fail; the conditions of failure are identifiable; and failure has an actual effect on subsequent generation. A generative rule that cannot fail, however formally complete, cannot correspond to the real world. This is the positive discipline into which the diagnosis of Part I is transformed in this part.

Under this discipline, this part unfolds gradually around three kinds of questions: the minimal definition of an event, namely under what conditions an event is allowed to appear; the criteria of closure and settlement, namely how generation ends rather than how it continues; and freezing and reproducibility, namely how stable structures appear rather than how they are assumed. Every chapter follows the same methodological constraint: it proposes no proposition that cannot be rejected. One further point must be emphasized again, because it defines the relationship between this part and the entire Energy Quantum Theory series: this part does not take any existing differential form as a premise. All concepts such as gradients, frequencies, flows, and scales, if they appear in this book, must be redefined as results of generation rather than starting points of generation. This book is therefore neither a summary of the preceding nine works nor their technical extension, but a restart from a logically earlier and more fragile starting point.

Part II promises neither unification, nor an explanation of all phenomena, nor even success. It promises only one thing: if generation fails, the failure will be explicitly marked; if generation succeeds, the success will occur within finite rules. In this sense, Part II is not a completed physical theory, but the minimal grammatical layer through which physical generation can become possible.

4. Events Rather Than Objects

Part I has cleared the starting point: one cannot begin from the container, from continuity, or from probability. Yet clearing the ground does not by itself constitute a theory. To begin again from the cleared starting point, one must first answer the most basic question: what is the minimal bearer of generation? The answer given in this chapter is: the event. But the word "event" has long been occupied by various meanings in physics. It is often equated with a spacetime point, a particle, a segment of a process, or an observational result. The work of this chapter is therefore first negative: it excludes, one by one, these four most common equivalences, which cannot hold within generative logic (4.1–4.4), and only then turns to the positive definition (4.5), ultimately elevating this definition from a convention into a starting point of argument (4.6). What this chapter seeks to establish is a generative starting point that depends neither on background, nor on objects, nor on observation.

4.1. An Event Is Not a Spacetime Point

In almost all existing physical theories, an "event" is by default understood as something that occurs at a certain spacetime location. Whether in classical mechanics, where "a particle has a certain state at (x,t)," in relativity, where one speaks of a "spacetime point-event," or in quantum theory, where "measurement occurs at a certain moment," an event is treated as a marker in spacetime. This understanding is not a conclusion, but a long-unexamined premise.

The problem is that, in standard physical formulations, "event =

spacetime point" appears to be a definition, but in fact performs a conceptual smuggling. The procedure is this: given a four-dimensional coordinate (t,x,y,z), assign some physical quantity or change to that point, and call this assignment an "event." Before "event" has been explained, this procedure has already presupposed at least three things: spacetime already exists as an available containing container, the event can be located, and location itself is a physically meaningful operation. Once these premises are accepted, the question of the existence of the event is quietly dissolved—the event no longer needs to be generated; it only needs to be labeled.

Here one must distinguish two fundamentally different kinds of things: mathematical points and generative events. A mathematical point is an element of a continuous manifold; it can be listed a priori, approached without limit, and has no relation to "occurrence" itself. A generative event, by contrast, is the result of an occurrence or a failure. A spacetime point belongs to the former. Once a generative event is reduced to a spacetime point, it immediately loses three key properties: indivisibility, the uncertainty of "whether it occurs," and the possibility of failure. In other words, "occurrence" degenerates into "a coordinate has been assigned a value." This degeneration also reverses the logical order. Traditional accounts habitually begin with "where and when it occurs," but in generative logic, the more basic question is not "where does it occur," but "whether something occurs"; only after "whether it occurs" has been adjudicated does "where it occurs" acquire meaning. The spacetime-point model gives the "where" first and thereby defaults to "it necessarily occurs." This inversion makes "event failure" semantically inexpressible.

Quantum theory has not escaped this structure, but has instead pushed it to an extreme. The so-called "measurement event" is usually formulated as the collapse of the wave function at a certain moment t and at a certain position x. Here the "event" is still a preassigned position–moment label, with collapse merely appended to it as an additional rule. This is one root of the persistent difficulty of the measurement problem: collapse is required to occur at an already existing spacetime point, and therefore can exist only as an additional postulate, rather than becoming a result of generation.

The task of this section is only exclusionary, so it does not yet provide the full positive definition of an event, but only states the necessary negative condition. If an "event" is to become the basic unit of a theory of generation, it must at least satisfy the following: its existence cannot be automatically guaranteed by coordinates, its occurrence must be capable of failure, and its structure cannot depend on the continuous limit. Any "event" that does not satisfy these conditions, however convenient in calculation, can serve only as a descriptive tool and not as an ontological unit. The conclusion of this section is therefore negative and strict: an event cannot be understood as a point in spacetime. This denial does not reject the validity of spacetime readout, but explicitly postpones it until after generation—only after an event is permitted to "occur or not occur" can spacetime readout appear as a derived structure.

4.2. An Event Is Not a Particle

After rejecting "event = spacetime point," the most common and most persistent alternative is to understand an event as a microscopic particle or its instantaneous state. This approach appears closer to physical intuition, but it likewise cannot hold within generative logic.

The intuitive appeal of the particle model is easy to understand: a particle can be counted, named, and tracked; it maintains its identity in time, occupies a position in space, and exchanges conserved quantities in interaction. This image of a "persistent identifiable object" makes the particle the most natural candidate for ontology. Yet it is precisely this persistence that makes the particle unsuitable as a unit of generation. A particle is presupposed as an already existing object that maintains itself through evolution; its existence is not the result of generation, but an assumed premise. In the particle picture, the question is always "how does the particle move and interact," never "whether the particle is generated."

One might retreat a step and say: what counts as an event is not the particle itself, but the interaction among particles. This does not solve the problem; it merely pushes the presupposition back one level. Interaction still presupposes that the particles participating in it already exist. What is taken as the "event" is still something that happens among already existing objects, not the generation of the objects themselves. Quantum field theory appears to move beyond the particle picture by reinterpreting particles as excitations of fields; yet the field is presupposed as an a priori existence pervading spacetime, and excitation is only the local behavior of the field against a given background. So-called "creation" and "annihilation" events still occur within an ungenerated field–spacetime container. The ontological role of the particle is replaced by the field, but the grammar in which "the object precedes generation" remains unchanged.

The fundamental limitation of the particle perspective is that it cannot express "generative failure." A particle either exists or does not exist, but its existence is never an attempt that might fail. The continuity of the worldline guarantees the persistence of the object and thereby erases the position of failure—in the particle picture, there is no structural room for "an attempt to generate that did not succeed." The order of expression must therefore be reversed: not "an object has undergone something," but "occurrence (generation) has produced a stable pattern that appears as an object." The object is not the starting point of generation, but the product frozen after events repeatedly hold. In the final analysis, a particle is a persistent identifiable object, whereas an event is a one-time generative adjudication; the former presupposes existence, while the latter bears existence. If the particle is taken as the ontological starting point, generation inevitably degenerates into state updating, and failure is inevitably erased by the continuity of the worldline. Therefore, an event cannot be a particle.

4.3. An Event Is Not a Continuous Segment of a Process

After excluding points and particles, a more refined alternative is to understand an event as a segment within a continuous process. This intuition is attractive because it seems both to preserve the dynamic sense of "occurrence" and to avoid rigidifying the event into a static point or object. But "process" is here made a priori: it is presupposed as an already flowing continuous trajectory, and the event is only a segment cut out from this trajectory. The existence of the process itself is never questioned, but is given by default. The problem of generation is again pushed onto an ungenerated background, except that this time the background is a "continuous process" rather than a "spacetime container."

More fundamentally, a continuous process cannot provide a boundary of "occurrence." A continuous trajectory can be divided and cut arbitrarily; it contains no intrinsic point at which one can judge "completed here, not completed there." Generation, however, requires precisely such a boundary—whether something has occurred must be a decidable fact, not an interval that can be divided arbitrarily. In the continuous-process picture, an "event" is often only a narrative label: for descriptive convenience, we mark certain "salient points" on the trajectory, such as peaks, inflection points, or phase transitions, but these marks do not correspond to any intrinsic generative completion. Their "eventhood" comes entirely from external description, not from the process itself. Like continuous grammar, the process model also systematically erases failure: on a continuously evolving trajectory, there is no place where "interruption means failure." Any stagnation is rewritten as "tending toward a limit," any rupture is handled through piecewise or weak solutions, and failure once again loses its ontological position.

One might appeal to phase transitions, abrupt changes, or bifurcations, and regard them as genuine "events" within a process. Yet these concepts still belong to the category of evolution: a phase transition is the result of continuous parameter change, and a bifurcation is a change in solution structure. They describe "how already generated structures change," not "why a structure is generated here." Their "eventhood" is pseudo-eventhood: a feature of evolution, not an adjudication of generation. We can therefore see the ontological conflict between generative events and continuous processes: a generative event requires an indivisible boundary of completion and the possibility of failure, whereas a continuous process is characterized precisely by infinite divisibility, absence of boundary, and incapacity for failure. The two are logically incompatible. This book therefore rejects the understanding of an event as a segment of a process and requires a retreat from the "narrative of evolution" to "generative adjudication": not asking "how the trajectory extends," but asking "whether a generative attempt closes within finite steps." A continuous process presupposes an already flowing trajectory, and can neither provide an intrinsic boundary of occurrence nor bear failure. To reduce an event to a segment of a process is to outsource the problem of generation to an ungenerated continuous background. Therefore, an event cannot be a continuous segment of a process.

4.4. An Event Is Not an Observational Result

The final common alternative is to equate an event with the result of observation or measurement—the "event" is the definite output recorded by an instrument. This position is operationally natural, but it implies a strong premise: the existence of an observer or measuring apparatus, whose intervention is the condition under which facts become established.

But a measurement result belongs to the readout layer, not to the generation layer. An instrument reads the projected output of an already generated structure under finite bandwidth and finite resolution. Readout records facts, but does not generate facts. Equating an event with readout transfers the authority of "occurrence" to the recording system. The collapse narrative of quantum measurement is the extreme form of this transfer: the fact is said to be "selected" at the moment measurement occurs. This constitutes a logical inversion. The fact should be produced by generation and recorded afterward by readout, but the order is reversed into readout producing the fact, and the responsibility for generation is wrongly assigned to observation.

If one insists that an event is an observational result, an unacceptable consequence follows: "no observation, no fact." This would mean that before any observer or measuring apparatus existed, the world had no generated facts—the evolution of the early universe and the formation of stars would become ontologically unsayable. This consequence alone is sufficient to reject the position. (In Part IV, this book will state explicitly that facts at the generation layer are objectively established before any readout, and that readout is only a posterior projection; see Chapter 14.) Observational results also lack genuine failure: measurement either gives a result or is not performed, but "giving a result" is never a generative attempt that might fail. Failure again has no place. This book therefore insists on a clear and irreversible order: generation precedes observation. Facts are first produced at the generation layer (they close and freeze), and only afterward can they be projected by the readout layer under finite protocols as observational results. Observation never constitutes facts; it only records them. Equating an event with measurement output transfers generative authority to the recording system, makes generative logic depend on external intervention, and leads to the unacceptable consequence of "no observation, no fact." Therefore, an event cannot be an observational result.

4.5. The Event Is the Unit of Generation

After completing this series of negative clarifications, this section turns to the positive definition: in this book, an event is defined as the minimal, irreducible unit of generation. This assertion is not a rhetorical choice of position, but is compelled by the minimal requirements of generative logic.

Any theory that takes generation as its starting point must first answer: what is the minimal bearer of generation? If this bearer is still an object, a state, probability, or an observational result, then generation necessarily depends on some a priori structure and loses its ontological independence. A qualified unit of generation must therefore satisfy at least four minimal conditions: it must not presuppose a background, meaning it must not depend on spacetime, fields, objects, or state spaces; it must not be continuously decomposable, meaning that there are no "smaller generative fragments"; it must have the possibility of failure, meaning that generation does not necessarily hold; and it must be decidable, meaning that whether generation occurs must be a logical fact. The only candidate that satisfies these conditions is the event.

Accordingly, in this book, an event is not defined as "what has happened," but as a generative adjudication; more precisely, an event is the minimal unit of adjudication concerning whether a given generative attempt holds. This adjudication contains three intrinsic possibilities: holding (successful generation), not holding (generative failure), and non-closure (generative suspension). The event itself carries no object, numerical value, or trajectory; it carries only one result: whether this generation is accepted as fact. Because an event concerns only generative adjudication, it is logically prior to all structural objects—prior to particles, fields, spacetime, and probability distributions. Structure is not the starting point of generation, but the stable product after events repeatedly hold; an object is not the bearer of events, but a frozen event pattern.

The exact meaning of "unit of generation" must be clarified: it does not mean that the event is the smallest physical quantity, but that the event is indivisible in generative logic. Any attempt to decompose an event into smaller time intervals, finer spatial points, or lower-order probabilistic steps would reintroduce a continuous background and thereby destroy the ontological status of generation. An event is not an "instant," but an indivisible act of adjudication. As a unit of generation, an event must also contain the possibility of failure. This distinction is fundamental: an object either exists or does not exist; a state either takes a value or updates; observation either yields a result or is performed. An event is different: an event may attempt generation, but does not necessarily succeed. Failure is not an anomaly, noise, or lack of knowledge; failure is an ontological component of generative logic. Correspondingly, an event carries no probability at the ontological layer: probability can appear only at the statistical readout layer of many event results, and cannot be used to describe whether a single event is generated. An event does not "occur with some probability"; it is either generated or not generated. Probability is a readout structure, not a generative mechanism.

Finally, an event as a unit of generation does not automatically form a temporal sequence. Without presupposing spacetime, there is no before-and-after order among events, but only generative dependence and closure relations. What is called "order" is the projection of a later structure. Therefore, this book discusses not an event chain, but an event network—a logical structure composed of generative adjudications and relations of failure. In sum, the conclusion of this section can be summarized in three points: an event is not a property of an object, but an adjudication of generation; an event is prior to all physical structures and is the minimal unit of generation; failure is not a defect, but an intrinsic possibility of the event.

4.6. The Event as the Minimal Condition: Four Criteria for How Generation Can Hold

Section 4.5 defined the event as the minimal unit of generation. But this definition might itself be challenged as a convention—one might ask: why take "generative adjudication" as the starting point rather than writing existence directly into the premise? To respond to this challenge, this section elevates "the event is the unit of generation" from a position into a starting point of argument. It does so by showing that any theory claiming to "explain how structured existence is achieved without presupposing spacetime" must satisfy four cumulative minimal conditions. These four conditions are not derived backward from a mechanism favored by this book, but are forced by the requirement of taking the task of generation seriously; they therefore do not belong exclusively to the framework of this book, but constrain any theory within the same problem domain.

The relation between this section and the structure of the entire book must be stated in advance, because it will be used repeatedly in later chapters. These four minimal conditions establish a hierarchy of requirements: what any pre-physical theory of generation must preserve. The closure and freezing that this book subsequently introduces are only one local implementation of these fourfold requirements, especially the fourth, and not the requirements themselves. Whenever the reader suspects that closure or freezing is "too specific" or asks whether it is unique, one should return to this section: the specific mechanism may be replaced, but the minimal condition it must fulfill cannot be avoided.

First minimal condition: structured existence must be a result within the process, not a premise. If a theory claims to explain how structured existence is achieved, it cannot already treat that existence as something obtained at the starting point of the explanation; otherwise, what it explains is not "achievement," but only displays later what was already tacitly permitted at the beginning. The key distinction here is not the narrative order of "appearing first/appearing later," but two different explanatory statuses: does the process bear the achievement of existence, or does it merely display a result already guaranteed by the starting point? A practical criterion is: if the process described by the theory does not unfold as described, is structured existence thereby not achieved? If the answer is no—if existence is already independently guaranteed by the starting point—then the process does not bear the achievement, but only displays an already existing result. The reason the event as generative adjudication can bear achievement is precisely that it contains the possibility of failure (4.5): every adjudication is a real advancement in the status of a candidate, not a rearrangement of something already given.

Second minimal condition: the achievement of failure must be internal to the same process, not externalized. Once structured existence is a real result within the process, its opposite—non-achievement—cannot be pushed entirely outside the process, nor can it be redescribed as mere "delay." If non-achievement is externalized or dissolved into an indefinite "not yet complete," then achievement itself loses the contrastive space that gives it meaning and collapses back into "tacit admissibility." This is why this book elevates failure into an ontological component and repeatedly rejects, in Part I, the deferral grammar of "always just short of completion" (see Chapter 2): failure must exist as an internal possibility of the same generative process.

Third minimal condition: the three statuses of achieved, undecided, and failed must be genuinely distinguished, and consequences must not be exempted indefinitely. A mere binary distinction between "achieved" and "not achieved" is insufficient. Taking process seriously requires distinguishing at least three statuses of a candidate: structured existence has been achieved; the process is still internally undecided; the process has failed to bring about structured existence. These three correspond to different theoretical roles and cannot collapse into one another. It is especially crucial that "undecided" cannot serve as an indefinite shelter, allowing "failure" never to appear as a real status of the candidate. If anything not yet achieved can always be redescribed as "still undecided," then the second condition is preserved in wording but emptied in practice. This book will fulfill this distinction through decidable closure criteria in Chapters 8 and 12: undecided means that differences are still accumulating but have not yet reached the threshold of closure; failure means that the accumulated trajectory no longer tends toward closure; achieved means that closure has already been established. The three are different accumulation histories, not different descriptions of the same ambiguous state.

Fourth minimal condition: there must be some internal threshold structure that makes these three statuses decidable by the process itself, rather than externally imposed classifications. The third condition establishes that the three statuses must not collapse, but it has not yet answered: by what does the process internally make these statuses decidable? If a status can only be assigned after the fact by an external observer, then structured existence is no longer decided within the generation layer, but degenerates into a posterior classificatory label. The fourth condition therefore requires that there be some internal threshold structure by which a candidate can be adjudicated as occupying a status and can pass from one status to another. Here "threshold structure" should first be understood functionally rather than ontologically—it need not be a sharp cutoff point or a single observable marker, but may be distributed stabilization, relational locking, or satisfaction of global constraints. The point is only that the assignment of status must not be floating, retrospective, or purely classificatory.

It is precisely at this fourth condition that the place of this book's core mechanisms can be situated exactly. The closure and freezing subsequently introduced by this book are one local implementation of the fourth minimal condition, not the condition itself, nor the only possible implementation. Closure is primarily a condition concerning phase differences: it marks the accumulation of phase differences converging to an internally decidable completed state. Freezing is primarily a condition concerning directional stabilization: it marks the accumulation of directional differences producing a stable relational organization, so that the closed state is retained rather than immediately returning to undecidedness. Neither is guaranteed in advance by the nature of the process—this is precisely what distinguishes them from final states of the "necessarily achieved" kind. They are results that the process may or may not reach through the actual accumulation of differences. Any competing theory may implement the fourth condition in different ways, such as distributed stabilization, relational locking, or the satisfaction of global constraints. What it cannot do is leave the adjudication of status entirely outside the process. This book chooses closure and freezing in order to concretely show what such an implementation must accomplish, not to monopolize the space of solutions.

These four conditions form a cumulative sequence: each is forced by the burden generated by the previous one. The first requires existence to be a result rather than a premise; the second requires failure to be internal to the same process; the third requires the three statuses to be genuinely distinguished and not indefinitely exempted; the fourth requires these statuses to have an internal threshold basis within the process. Satisfying the preceding condition does not automatically satisfy the next, and without the final step, the first three remain conceptually valid but structurally suspended. The remaining chapters of Part II—triadic intrinsic structure (Chapter 5), relation (Chapter 6), configuration and logical time (Chapter 7), phase evolution and closure (Chapter 8)—can all be read as the gradual fulfillment of these four requirements. Together, they explain how the event as unit of generation can, without presupposing any background, make structured existence into a result that is decidable within the process, capable of failure, and capable of being retained.

4.7. Summary of This Chapter: The Event as the Only Starting Point of Generation

The task of Chapter 4 has not been to propose a new physical object, but to complete a more fundamental ontological clearing: to detach the status of "occurrence" completely from objects, spacetime, and observation.

Through a series of negative clarifications, this chapter has systematically excluded three of the most common ways of equating events in physical theory, each of which fails within generative logic. First, an event is not a spacetime point: equating an event with a marker in coordinates in effect presupposes a prior container, making the event unnecessary to generate and incapable of failure; "occurrence" is quietly replaced by "being located." Second, an event is not a particle or an interaction: a particle is an object with persistent identity, whereas an event is a one-time generative adjudication. Once the particle is taken as the ontological starting point, generation degenerates into state updating, and failure is erased by the continuity of the worldline. Third, an event is not an observational result: observation always occurs after generation; it reads facts but does not generate facts. Equating an event with measurement output transfers generative authority to the recording system and makes generative logic depend on external intervention.

After completing these exclusions, this chapter gave its positive conclusion: an event is the minimal, irreducible unit of generation. In this book, an event carries no object, numerical value, or trajectory, but only the most basic adjudication—whether a given generative attempt holds. For precisely this reason, the event is logically prior to all physical structures: prior to particles, fields, spacetime, and probability. The indivisibility of the event ensures that generation is not diluted by continuous grammar; the possibility of failure contained in the event ensures that the theory bears real risk; and the non-probabilistic character of the event strictly confines probability to the readout layer rather than the ontological layer.

Finally, this chapter elevated "the event is the unit of generation" from a convention into a starting point of argument (4.6): any theory claiming to explain the achievement of structured existence under background-free conditions must satisfy four cumulative minimal conditions—existence must be a result within the process, failure must be internal to the process, the three statuses must be genuinely distinguished and not indefinitely exempted, and there must be an internal threshold structure that makes status decidable. The closure and freezing that this book later introduces are only one local implementation of the fourth condition, not the requirement itself. This framework both provides a testable argumentative skeleton for subsequent chapters and anticipates the objection that the "unit of generation" is merely a convention. At this point, the chapter has established a generative starting point that depends neither on background, nor on objects, nor on observation. Yet the event as unit of generation is still not enough to constitute a world; only when events can bear some intrinsic structure, and thereby enter closure and freezing, can "occurrence" be transformed into "fact." Therefore, the next chapter will no longer ask "what is an event," but will turn to a more crucial question: what is the intrinsic structure of the event?

5. The Triadic Intrinsic Structure of the Event

If the event is the minimal unit of generation (Chapter 4), then it must carry some intrinsic structure that allows generative attempts to be compared, closed, and frozen. This chapter provides that structure.

Before discussing each component, however, we must first correct an almost unavoidable misunderstanding that will otherwise run throughout the chapter: the misunderstanding that the intrinsic structure of the event consists of three parallel and independent attributes—as if frequency, phase, and direction were three separate features, each requiring separate justification and accidentally assembled into a set. This understanding appears harmless, but it plants a problem at the conceptual root. Once direction is treated as an "independent spatial orientation" parallel to rhythm, it again presupposes the space toward which it points, and a geometric circle thereby arises. Once the three are treated as equal and independent factors, the product structure $g \cdot k_\theta \cdot k_d$ in the weight kernel will be misread as "the multiplication of three independent factors," thereby concealing that they are different aspects of the same thing.

This chapter therefore begins from a unified conceptual structure: rhythm is the core; phase is the state of rhythm; direction is the extension direction of rhythm—more precisely, the tendency of an event to extend intrinsically toward certain relations. The three are not parallel independent attributes, but three dimensions of a unified concept of event, jointly constituting "what the event is." Section 5.1 establishes rhythm as the core; 5.2 shows how phase, as the internal state of rhythm, derives from rhythm; 5.3 shows how direction, as the tendency of rhythm toward relational extension, derives from rhythm; 5.4 argues that this three-dimensional structure is irreducible and requires nothing further; 5.5 turns to the generation layer and explains how this unified structure at the ontological layer becomes operational through differences.

5.1. Rhythm: The Core Specification of the Event

As a unit of generation, the most central and first-established intrinsic specification of an event is rhythm—a rhythmic processual existence. In the formalization of this book, rhythm is represented by the frequency scalar f; but it must be emphasized from the outset that frequency is here the measure of rhythm, while rhythm is what is being specified. Phase and direction will both be derived from this core as two dimensions of rhythm itself, rather than as two additional items parallel to it.

The "frequency" discussed here is not a time function in the usual sense. It does not mean the number of oscillations per unit time, a period measured relative to an external clock, or a rate of change depending on a continuous time parameter. In a generative framework with no a priori spacetime, there is no temporal background on which events can "occur," and therefore no "frequency that changes with time." Rhythm is not a dynamical quantity, but a rhythmic scale carried by the event itself—it gives the most basic characterization of the event as processual existence. More precisely, rhythm is the scale of generation. Whether an event can be adjudicated as generated successfully or as failed is not arbitrary, but depends on the matching relation between the event's internal generative intensity and its generative environment; this matching must rely on some scale, otherwise "whether closure occurs" cannot be adjudicated. Rhythm is the minimal realization of this scale: it characterizes not "how fast" an event occurs, but the density and compactness with which the event attempts to occur at the generation layer; it marks the intensity of a generative attempt in logical depth, not a rate of motion along a temporal axis.

This introduces a fundamental reversal. In traditional physics, frequency is always defined by time, $f = 1/T$; in generative logic, the relation is precisely reversed—rhythm is the original intrinsic property of the event, and time is a derived readout structure formed after the rhythmic relations among many events have frozen. Events do not oscillate in time; time emerges from the rhythmic structure of events. This reversal is not rhetorical, but structural. It marks the fundamental divergence between this book and all theories that begin from a differentiable time parameter.

The core function of rhythm lies in its serving as the basis of the compatibility criterion, not as a dynamical factor. Not all events can close with one another. Closure requires strict conditions, the most basic of which is rhythmic compatibility: if the gap between rhythms is too large, no stable closure can form; if rhythms are close or proportional, a rhythmic closure loop is allowed; if rhythms are completely mismatched, the generative attempt fails. Rhythm therefore does not determine "what happens," but determines whether any sustainable generative structure can occur. The wording here must be noted carefully: what has adjudicative force is the gap, proportion, and mismatch among rhythms—that is, rhythmic difference, not the absolute value of any single event's rhythm. This point will be systematically established in Section 5.5 as the Difference Principle.

Two strict distinctions must also be made to avoid conceptual misplacement. First, rhythm is not energy, but it can be read as energy: at the ontological layer, rhythm describes generative rhythm, not an exchanged quantity, a conserved quantity, or a measure of work; in later readout protocols, rhythm can be mapped onto an energy scale because the two have a readable correspondence in their role as "generative density," but this correspondence belongs to a choice at the readout layer, not to a definition at the generation layer—readout equivalence must not be mistaken for ontological identity. Second, rhythm is not probability of failure: rhythm is the intrinsic generative scale of the event, while probability is a statistical description at the readout layer concerning the results of many events. The two are logically distinct; a single event can only generate or fail, and there is no ontological statement that it "generates with some probability." Finally, it must be noted that the event world does not contain a single dominant rhythm, but naturally presents a structure of multiple coexisting rhythms—high-rhythm events have high generative density, high closure cost, and strong locality; medium-rhythm events most readily form stable structures capable of freezing; low-rhythm events generate sparsely but have broad relational reach. This multi-rhythmic structure is not a posterior classification, but a logically natural result once events are understood as units of generation. In sum, rhythm is the intrinsic generative scale of the event rather than a time function; it precedes time and is the precondition for closure and structural formation; it provides a compatibility criterion rather than a dynamical law. The next two sections will show that phase and direction are not two additional items parallel to rhythm, but two indispensable dimensions of rhythm itself.

5.2. Phase: The Internal State of Rhythm

Once rhythm has been established as the core, the first derived dimension is phase. If rhythm answers the question "with what rhythm does the event attempt generation," then phase answers a more adjudicative question: in which internal state of its rhythm is the event now? The key point is that phase is not an independent attribute parallel to rhythm, but the state of rhythm itself—the position rhythm occupies in the unfolding of its own sequence. Without rhythm, there is no phase; phase is the state of a rhythm at a certain moment of its recursion. Under the conditions of no a priori spacetime and no continuous evolution parameter, this "state at this moment" cannot be expressed through a "temporal position," but only introduced as the internal state of rhythm.

A nearly unavoidable misunderstanding must therefore first be excluded: the phase discussed here is not phase in traditional wave theory. It does not mean an angular parameter in a sine or exponential function, a periodic position relative to continuous time, or a geometric marker in a propagating medium. Phase does not depend on a wave equation, nor on any external propagation structure. It is the internal state marker of the event within its own rhythmic cycle, the internal moment of rhythm, not another quantity outside rhythm. The physical meaning of this internal state is generative progress. Even at the minimal event level, generation is not completed instantaneously. A generative attempt includes at least several logical stages: not yet entering the generative window, entering the settlement interval, successful closure, or failure and exit. Phase characterizes precisely the progress of the event within this internal generative process—when phases are the same, generative states can align; when phases are offset, generative states cannot superpose; when phases are completely mismatched, no closure can form. Phase is a necessary condition for generative coordination precisely because it is the state of rhythm itself: without it, rhythm would become a process without internal differentiated moments, and therefore not any determinate process.

The ontological priority of phase is also expressed in its priority over causal order. In traditional theories, causal order is usually understood as a direct reflection of temporal order. In generative logic, this relation is completely reversed—at the ontological layer, phase relations precede any definable causal order. Only when multiple events form stable phase relations can structures such as "before–after" and "cause–effect" be frozen as recognizable orders at the readout layer. Causality is not a basic structure, but a semantic projection of phase relations after freezing.

Here the first prefiguration of the Difference Principle appears, and it is decisive for the entire book: at the layer of generative ontology, there is no reference for absolute phase. The phase of a single event, as the internal state of its rhythm, is its intrinsic specification at the ontological layer, but it cannot itself be externally observed and enters no generative operation; only phase differences among events have adjudicative significance, determining whether generation can coordinate, repel, or fail. In other words, the absolute value of phase is the intrinsic specification of the event (ontological layer), while its only operational form is phase difference (generation layer). This blocks from the outset any metaphysical assumption of a "global generative clock" or "universal unified phase"—phase operates relationally from the beginning, not through absolute values. It is precisely phase difference that constitutes the core criterion of closure: closure does not mean that events "happen to occur simultaneously," but that multiple generative attempts complete a consistent settlement at the level of state. If phase differences fall within the admissible interval, closure is possible; if phase differences cross the threshold, closure fails; if phase relations repeat stably, they can freeze into a factual structure. Closure is therefore not geometric coincidence, but the consistent completion of generative states.

Phase also reveals one source of irreversibility. Once closure is completed, the corresponding phase relation does not return to the "ungenerated" state. This irreversibility does not arise from external dissipation, but is a structural result of generative logic itself—the advancement of phase constitutes part of generative history; once settlement occurs, history is written in, and history is not revocable. We can therefore see that the root of the arrow of time lies not in entropy, but in the irreversible completion of phase settlement. (This irreversibility will be concentratedly elevated in Section 7.4 and Chapter 13 into "freezing monotonicity," serving as a unified argument for the arrow of time.) Finally, the same distinction made for rhythm must be made here: phase belongs to the layer of generative ontology, while probability belongs to the statistical readout layer. Phase does not describe "the possibility of occurrence," but the "state of generation." A single event either closes successfully or fails to generate; there is no ontological expression in which it is "generated with some phase probability." In sum, phase is the state variable of rhythm, not a wave or temporal parameter, and not an independent attribute parallel to rhythm. Phase difference, rather than absolute phase, determines whether events can coordinate in closure; causal order and the arrow of time arise from the freezing of phase relations.

5.3. Direction: The Tendency of Rhythm Toward Relational Extension

After rhythm has been established as the core and phase as the internal state of rhythm, the second derived dimension is direction. Even if events are rhythmically compatible and phase-aligned, one problem remains unresolved: how can they form distinguishable, extendable, and stable relational structures? Without this capacity, generation could remain only at the level of adjudicating "whether closure occurs," but could not produce any factual structure with internal differentiation. Direction is the third dimension introduced for this purpose. Yet like phase, direction is not an independent attribute parallel to rhythm, but the extension direction of rhythm itself—the intrinsic tendency of rhythm toward certain relational extensions.

This formulation must be stated positively, rather than defensively as "not spatial direction": direction is the extension direction of rhythm, more specifically the relational direction of the event—the intrinsic tendency of rhythm toward certain relations. It exists at the ontological layer as a constitutive dimension of rhythm, while its realization as extension belongs to the generation layer. Direction is not the event's orientation toward external space, but the event's orientational tendency toward other events, and therefore it fundamentally presupposes no background geometry. This understanding cuts off geometric circularity at once: if direction is taken as an "independent spatial orientation," it presupposes the space toward which it points; but once direction is correctly understood as rhythm's intrinsic tendency to extend toward certain relations, it does not point to any spatial position at all, and geometric circularity cannot arise at the conceptual root. Direction is the event's intrinsic bias toward "connectability"—that is, the tendency of rhythm to extend toward certain kinds of relations—not a geometric property of an object in space.

The function of direction is to serve as an orientational bias of relation. Events are not equivalently open to all other events. Even when rhythm allows compatibility and phase allows alignment, relation still requires selectivity. Direction characterizes precisely this selectivity—which relations are more easily reinforced, which are more easily suppressed, and which generative paths are more likely to be repeatedly attempted. It is not an added label, but the minimal source of asymmetry in the formation of a relation network. This "orientation" is always relational: it says "toward what kind of relation this rhythm tends to extend," not "in which spatial direction this event lies." It is this asymmetry that makes geometry possible. If events carried only rhythm and phase, all closeable relations would remain logically symmetric. Closure could occur, but structure could not be differentiated, and there would be no internal morphological difference. The introduction of direction breaks this degeneration—relations begin to have orientational differences, multi-event closure begins to display internal structure, and geometry thereby ceases to be an a priori existence and becomes the accumulated result of directional biases within a generative network. This also requires reversing the usual order of spatial dimension: it is not that dimensions come first and directions afterward, but that directional relation networks come first and structures readable as "dimensions" come afterward. Dimension is not an intrinsic label of space, but the readout result of many frozen directional relations. This allows "geometry" to acquire generative meaning without presupposing a manifold.

Direction also participates in the selection of stability. Not all directional configurations are stable: in multi-event closure, some directional combinations reinforce relational loops, some introduce internal tension, and some lead repeatedly to closure failure. Direction therefore not only generates geometry, but also participates in selecting which structures can freeze over the long term. Stable structures are not generated arbitrarily, but are the result of repeated elimination among success and failure in directional configurations. A strict distinction must be made here: direction is not the "direction of force." Force belongs to the explanatory language of the readout layer, while direction belongs to relational bias at the generation layer. At the generation layer, there are no objects being pushed, only relational paths being reinforced or weakened. Misreading direction as force is precisely a residue of object–container grammar within generative theory.

Direction also has a key structural difference relative to phase, and it must be stated here because it is one of the supports for the later argument concerning irreversibility: direction is the extension tendency of rhythm, a fixed intrinsic property of the event before freezing, and is not updated by phase dynamics. Phase evolves according to update rules at each logical step, while direction does not change under phase dynamics. This fixed character of direction provides a stable intrinsic basis for freezing—when a candidate structure freezes through a stable directional organization, that organization is retained not by continuous dynamical adjustment, but by the fixed directional tendencies of the component events. (This point plays a crucial role in Chapter 13, in the discussion of freezing and irreversibility, and in the argument for freezing monotonicity: the fixed character of direction is one structural source of the irreversibility of freezing.) Finally, the introduction of direction necessarily brings with it the possibility of failure: even if phases align and rhythms are compatible, directional configurations may still prevent relations from closing, cause tension to accumulate, and force generative paths to terminate. This failure is not a defect, but an indispensable mechanism of structural selection. In sum, direction is the intrinsic tendency of rhythm to extend toward certain relations, the minimal intrinsic bias of relational orientation rather than a spatial coordinate; geometry emerges from directional relations rather than from a presupposed background; the fixed character of direction provides a stable basis for freezing; and stable structures arise from the repeated selection of directional configurations in generation. At this point, the triadic intrinsic structure of the event takes unified form at the ontological layer: rhythm is the core, phase is the internal state of rhythm, and direction is the tendency of rhythm toward relational extension. The three are not three parallel attributes, but three dimensions of the same concept of rhythm, jointly constituting the minimal complete structure required for the event as a unit of generation.

5.4. Why These Three Dimensions Are Irreducible and Require Nothing More

The preceding three sections established rhythm as the core, with phase and direction as its two dimensions. We must now answer a decisive question: why does the intrinsic structure of the event consist exactly of these three dimensions—neither fewer nor more? This is not a classificatory question or a stylistic choice, but a question of whether generative logic can hold coherently. This section appeals neither to intuition nor to historical convention, but employs a strict completeness argument by exclusion. It should be stated in advance that this argument does not seek to prove that "three independent attributes happen to form a complete set," but rather that "rhythm and its two dimensions constitute an indivisible unified structure, and if any dimension is removed, rhythm can no longer operate."

The criterion of completeness must first be established. In this book, "completeness" does not mean "explaining everything," but that a generative framework must at least satisfy four minimal requirements simultaneously: generability—events must be able to form nontrivial generative attempts, rather than stand in static juxtaposition; closure—the generative process must have an endogenous criterion of completion, rather than an externally imposed truncation; freezing—some completed results must be stably retained and participate in subsequent generation; failure—not all generative attempts necessarily succeed, and failure must be a structural possibility. Any structure lacking any one of these capacities is not qualified to become the ontology of physical generation. These four requirements correspond to the four minimal conditions established in Section 4.6: generability corresponds to "existence must be a result within the process," failure corresponds to "failure must be internal to the process," while closure and freezing jointly correspond to "an internal threshold structure makes status decidable."

Let us now examine the consequences of lacking any one dimension. Rhythm alone: rhythm characterizes the intensity of a generative attempt, but does not distinguish generative stages, nor the orientation of relations. The direct consequence is that there is no way to adjudicate whether generation has been completed, no way to distinguish "in generation" from "already settled," and all compatible relations are logically equivalent. Such a system can only describe distribution and density, but cannot describe history, progress, or adjudication. What it lacks is precisely rhythm's own internal state, namely phase. Rhythm plus phase: once phase is introduced—that is, once rhythm has an internal state—the system can distinguish generative stages, provide closure criteria, and define "completed" and "not completed." But an irreparable defect remains: all closeable relations remain structurally symmetric, different closures have no internal criterion of differentiation, and one cannot explain why some structures freeze while others do not. Without direction, that is, without rhythm's tendency toward extension, closure can only generate "equivalent facts," not "structural facts." Generation remains at the level of adjudication and has not yet entered the level of geometry and morphology. Rhythm plus direction: if phase is skipped and only rhythm and direction are retained, rhythm provides scale and direction introduces relational bias, so complex networks can appear formally; but there is no way to adjudicate whether generation has completed, no way to distinguish transients from facts, and so-called "freezing" can only depend on external rules. The reason is that lacking phase means lacking the internal state of rhythm, and the criterion of completion is precisely a criterion concerning this internal state. Generation here becomes an indefinitely extending or arbitrarily interrupted process, closure is completely outsourced, and failure is no longer an endogenous result. Phase plus direction: if rhythm is absent and only phase and direction are retained, the problem is most fundamental—phase and direction are two dimensions of rhythm; once rhythm is removed, they lose what they are. Phase loses its rhythmic basis (it is the state of a rhythm), and direction loses its reference of intensity (it is the extension tendency of a rhythm). Closure conditions cannot be quantified or compared, and "completion" can only be a formal adjudication rather than an executable, repeatable, and selectable generative event. Without rhythm, there is no operational core of generation, and generation degenerates into a logical relation rather than a physical process. This case most clearly shows that phase and direction are not attributes that can exist independently of rhythm, but dimensions of rhythm.

Only when rhythm and its two dimensions are present together does generative logic first satisfy all the conditions of completeness: rhythm provides the core scale of generation and level differentiation; phase, as the internal state of rhythm, provides generative state and the criterion of completion; direction, as the extension tendency of rhythm, provides relational orientation and structural selection. Together, the three constitute an irreducible generative loop—but the true meaning of this "irreducibility" is that they are three dimensions of the same concept of rhythm, not three separable independent parts. If any dimension is removed, at least one condition of completeness fails, and rhythm itself can no longer operate; any additional attribute can only be a readout-layer or derived structure. This three-dimensional structure therefore has two decisive features. The first is indivisibility: phase and direction cannot exist apart from rhythm, because they are dimensions of rhythm, while rhythm cannot operate apart from these two dimensions, because without phase there is no state and without direction there is no extension. Therefore no dimension can be derived from or replaced by the other two—but this "non-derivability" is not independence among three separate things; it is the mutual dependence of three aspects within a unified structure. The second is generative sufficiency: all subsequent physical concepts—time, space, object, probability, force, and energy—can be rewritten as frozen readouts of these three dimensions. This is therefore not a "convenient choice," but the complete unfolding of the concept of rhythm under minimal constraints. In short: "rhythm–phase–direction" is not a set of three independently useful attributes that happen to fit together, but the indivisible three dimensions of a single concept of rhythm. Any structure with fewer than these three dimensions cannot simultaneously support generation, closure, freezing, and failure, and rhythm itself cannot operate; any attribute beyond these three can only be derived, read out, or restated. The ontological starting point of generation is thereby strictly fixed—not as three parallel postulates, but as one unified core concept and its complete unfolding.

5.5. Difference: How the Triadic Structure Operates at the Generation Layer

The preceding four sections established the intrinsic structure of the event at the ontological layer: rhythm and its two dimensions are the absolute and constitutive specifications of the event. Yet one question remains unanswered, and it is decisive for the entire framework: how do these absolute intrinsic specifications become operational quantities capable of driving generation? This section provides the answer and thereby precisely delineates the division of labor between the ontological layer and the generation layer.

The core of the answer is one principle: absolute quantities are inoperative; differences alone are operative. The ontological layer concerns "what something is," while the generation layer concerns "how it is compared." The rhythm, phase, and direction of a single event are intrinsic and absolute—but absolute quantities themselves are inoperative. This point must be established directly: the absolute phase θ_i of an event enters no update rule; what enters phase evolution is the phase difference $\theta_j - \theta_i$. The absolute direction $\mathbf{d}_i$ of an event does not by itself determine any coupling; what enters directional compatibility is directional difference, that is, the relative orientation between two extension tendencies. The absolute rhythm f_i of an event likewise produces matching only when compared with another f_j; what enters the compatibility kernel is rhythmic difference or rhythmic ratio. Therefore, the ontological layer specifies the triad as the constitutive content of the event—here rhythm, phase, and direction are intrinsic and absolute, do not depend on any external reference frame, and are existential specifications of the event itself. The generation layer appears only through comparison among events—rhythmic difference, phase difference, and directional difference become meaningful operational quantities only when two or more events encounter one another, and they form the operational basis of coupling, closure, and freezing.

This principle has three decisive consequences.

First, it realizes background independence, and does so mechanically rather than declaratively. Background independence is not the negative avoidance that "the ontological layer does not use coordinates," but the positive constraint that "the generation layer can operate only through differences." Differences naturally eliminate any shared external reference zero: a global translation of absolute phases or a global rotation of absolute directions automatically cancels out in operations on differences and changes none of the adjudications concerning coupling, closure, or freezing. This is why all generative operations in this book can be rewritten in purely relational form without relying on any external reference frame—not because such a rewriting is found after the fact, but because the generation layer from the beginning operates only on differences, and absolute quantities never enter update rules. (A formal verification of this purely difference-based rewriting will be given in Appendix B as a positive confirmation of background independence.) This also clarifies the apparent tension between absoluteness at the ontological layer and background independence: the two are not contradictory. The absoluteness of the triad is a constitutive specification in the ontological sense, the existential content that makes the event what it is; the requirement of background independence applies to the operational constraint at the generation layer, namely that only differences can be used. The former is "what the event is," the latter is "how events are compared." They belong to two different layers and perform different functions.

Second, it completes the final severance of the geometric circle at the operational level. Section 5.3 already cut the geometric circle conceptually—direction is the tendency of rhythm toward relational extension and does not point to any space. The Difference Principle completes the final link of this severance at the operational level. Consider the directional compatibility kernel: what enters it is never the absolute direction $\mathbf{d}_i$, but directional difference, namely the relative orientation between two extension tendencies (formally expressed as the relative compatibility $\mathbf{d}_i \cdot \mathbf{d}_j$); and a global rotation of absolute directional coordinates changes no directional difference and therefore changes no value of k_d. In other words, even if coordinates are used for notational convenience in formalization, those coordinates are merely a bookkeeping device for differences and can be globally rotated arbitrarily without affecting any generative adjudication. The directional kernel reads directional difference, not absolute direction—this is a positive statement derived from the Difference Principle, not an after-the-fact defense of "using a dot product." (When Appendices E and F use Euclidean coordinates and dot products in numerical implementation, they will explicitly state on this basis that coordinates are only a representation at the implementation layer: what they measure is directional difference, and they do not presuppose an external geometric container.)

Third, it provides the reason why logical steps belong to the generation layer. Difference becomes an operative quantity only when events encounter one another, and "encounter–comparison–update" itself constitutes a beat: each logical step is precisely one advance of an operation on differences. Therefore, logical steps count the number of advances in operations on differences. They intrinsically belong to the generation layer, not the ontological layer—the ontological layer consists of specifications, and specifications require no steps; the generation layer consists of advancing operations on differences, and necessarily has steps. (This point will be fully developed in Section 7.2 in the argument concerning logical time, and will be used there to respond to the objection that "logical steps secretly introduce temporal order.")

This chapter has now completed a full two-layer picture. At the ontological layer (5.1–5.4), the intrinsic structure of the event is rhythm and its two dimensions: rhythm is the core, phase is its internal state, and direction is its tendency toward relational extension. The three are the three dimensions of the same unified concept, absolute and constitutive. At the generation layer (5.5), the only operational form of this unified structure is difference: rhythmic difference, phase difference, and directional difference become operational quantities only when events encounter one another, and constitute the entire basis of coupling, closure, and freezing. The Difference Principle simultaneously realizes background independence, cuts off geometric circularity, and establishes the layer to which logical steps belong. The event is thereby no longer an abstract placeholder, but an entity that carries a unified intrinsic structure and enters generation through differences. The next chapter will further show that, precisely because the generation layer uses only differences, relation is not something externally attached to events, but the original form of operation at the generation layer.

5.6. Summary of This Chapter

This chapter has completed the most crucial step in generative ontology, and its completion involves a fundamental conceptual reset: it no longer expresses the intrinsic structure of the event as a stipulation of three parallel attributes, but as the three-dimensional internal unfolding of a unified concept of rhythm, while also completing the lower half of the picture in which this structure operates through differences at the generation layer.

At the ontological layer, this chapter established the following: rhythm is the core specification of the event, a rhythmic processual existence; phase is the internal state of rhythm, the position rhythm currently occupies within its own sequence; direction is the tendency of rhythm toward relational extension, namely the intrinsic tendency of rhythm to extend toward certain relations. The three are not equally independent attributes requiring separate justification, but the indivisible three dimensions of the same concept of event. This reset has three benefits. First, the argument is tightened from three separate defenses into the unfolding of one unified concept, and the force of the argument is thereby strengthened. Second, direction, understood as rhythm's tendency toward relational extension, cuts off geometric circularity at the conceptual root—it points to no space, and therefore presupposes no space toward which it points. Third, the product structure of the weight kernel is thereby correctly understood as the integration of rhythmic comparison across three dimensions, not as the multiplication of three independent factors. Through a completeness argument by exclusion, this chapter showed that if any one dimension is absent, at least one of the minimal physical requirements—generability, closure, freezing, and failure—fails; more importantly, rhythm itself can no longer operate, because phase and direction are themselves dimensions of rhythm. Remove either one, and rhythm becomes incomplete. Therefore, "rhythm–phase–direction" is not a modeling choice, but the complete unfolding of the concept of rhythm under minimal constraints.

At the generation layer, Section 5.5 completed the crucial lower half of the picture: the absolute intrinsic specifications of the ontological layer are themselves inoperative. Only differences—rhythmic differences, phase differences, and directional differences—become operational quantities when events encounter one another, and they constitute the entire basis of coupling, closure, and freezing. The Difference Principle fixes three things at once: it realizes background independence through a mechanism (differences eliminate external reference origins), it completes the final severance of geometric circularity at the operational level (generative operations read only directional differences, not absolute directions), and it explains why logical steps belong to the generation layer (the beat of operations on differences is the logical step). At this point, the event becomes an entity that carries a unified intrinsic structure and enters generation through differences. The minimal grammar of generative physics achieves its first coherent closure here—not as the union of three parallel postulates, but as a unified core concept, its two dimensions, and their only mode of operation at the generation layer.

6. Relations Are Not Additions

After establishing the triadic intrinsic structure of events (Chapter 5), this chapter turns to an equally fundamental question: what are relations between events?

Before entering the individual arguments, the overall position of this chapter must first be stated. It directly follows from the Difference Principle in Section 5.5. Chapter 5 has already shown that the intrinsic structure of an event—rhythm and its two dimensions—is absolute at the ontological layer, but its only mode of operation is through differences: rhythmic differences, phase differences, and directional differences become operational quantities only when two or more events encounter one another. This has a direct and decisive consequence for the ontological status of "relation": since all operations at the generation layer are carried by differences, and since a difference is by definition a quantity between two events, relation is not something additional attached to events, but the original form of operation at the generation layer. Relation is not a bridge constructed between isolated events after they first exist; on the contrary, the generation layer from the outset handles only differences between events, and "relation" is precisely the collective name for the organization, reinforcement, and selection of these differences. All the arguments in this chapter can be read as developments of this position from different sides: relation is not force (6.1), not probability (6.2), but generative reachability selected by operations on differences (6.6).

6.1. Weight Is Not Force

In the transition from object ontology to event ontology, one of the most common and most damaging misunderstandings is this: once "relational weight" is introduced, it is naturally understood as some kind of "force" or "interaction strength." This identification is almost unavoidable at the intuitive level, because modern physics has long compressed "relation" into "the action of force." But in generative logic, this identification is not an improper approximation, but a category mistake.

To see this clearly, we must first examine the definitional premises of force. In traditional physics, "force" implies at least three premises: there exists a measurable spatial structure, there exists a parametrizable temporal evolution, and there exist state variables that can be written into equations of motion. Force is defined as the causal term that produces changes of state, and its physical meaning ultimately becomes "a quantity that causes acceleration." But at the stage discussed by generative ontology, there is as yet no a priori space, no continuous time, and no object capable of acceleration—therefore all the definitional premises of "force" fail here. Weight does not depend on these premises: it does not measure pushing, but measures whether a generative relation is reachable. This is the ontological status of weight—it is generative reachability, not a pushing quantity. It answers not "how one event is pushed by another event," but "whether two events can still enter the same closure attempt"; this is an adjudication concerning generative permission, not a measure concerning the cause of motion.

The language of "force" arises automatically because of the deep inertia of container grammar: once we intuitively reintroduce space and time, the strength of a relation is immediately translated into the "strength of pushing or pulling." This is precisely the residue of container grammar in the concept of relation diagnosed in Chapter 1; identifying and suspending this inertia is the precondition for correctly understanding weight. Once it is suspended, the true meaning of changes in weight becomes clear: a change in weight is not motion, but a change in qualification. When w_{ij} increases or decreases, what changes is not the position or velocity of any object, but the qualification of whether "this relation is still allowed to enter the next generative settlement." A change in qualification is discrete, can be reset to zero, and is non-differentiable; this forms an ontological contrast with the continuity of motion. Conversely, if weight is equated with force, three consequences inevitably follow: a continuous background is reintroduced to carry the "force field"; generative selection is misread as dynamical evolution; and failure, namely the clearing of weight to zero, is misunderstood as an asymptotic process in which "force tends to zero" rather than as a discrete generative elimination. These three consequences jointly pull generative logic back into container grammar. If one must give weight an intuitive analogue, it is closer to "gating" or "channel openness": whether a relational channel remains open under the current generative state, rather than a force that pushes an object into motion. But even this analogy must be used cautiously outside the readout layer. This section therefore establishes: force is an explanatory language of the readout layer, while weight is a reachability criterion at the generation layer; relation precedes force—force appears only among frozen structures and only as an effective description at the readout layer.

6.2. Weight Is Not Probability

The second equally deep-rooted misreading is to treat weight as probability. This misreading must be excluded just as thoroughly, because it would drag generative logic—just rescued from the ontologization of probability (Chapter 3)—back into the same problem at the level of relation.

The fundamental obstacle is that the definitional premises of probability do not exist at the stage of generation. The definition of probability presupposes at least a complete sample space of events, a mutually exclusive and exhaustive set of outcomes, and repeatable sampling. At the stage of generation, none of these premises exists—there is no pregiven sample space, no exhaustive outcome set, and a single generation cannot be "sampled repeatedly." Therefore weight cannot be probability. From the questions they answer, the two are also categorically different: probability answers "with what frequency does a certain outcome appear," while weight answers "whether this relation is still allowed to enter the next generative settlement"; the former is a statistical description, while the latter is generative permission, and they are quantities of different categories. From the perspective of axioms, although weight takes values in $[0,1]$, it satisfies none of the basic axioms of probability: it does not require normalization ($\sum_j w(e_i, e_j) \neq 1$), does not require a complete event set, does not require mutual exclusivity, and does not require repeatable sampling; the meaning of $w = 0$ is "this relation no longer participates in any generation," not "an extremely small probability." The misreading of weight as probability originates from the historical inertia in quantum theory that treats quantities in the interval $[0,1]$ by default as probability amplitudes or probabilities; but the fact that a quantity takes values in $[0,1]$ does not make it probability. This is only a convenience of normalization, not the essence of probability.

Furthermore, the update mechanism of weight is not probabilistic updating: weight is selected according to the compatibility of the triadic intrinsic structure, namely differences, rather than revised according to Bayesian accumulation of evidence; it is generative selection, not probabilistic updating. If weight is interpreted as probability, the immediate consequence is that generative failure is reabsorbed as a "low-probability event," and therefore loses its ontological status. This is precisely the recurrence, at the level of relation, of the trap of probability ontologization diagnosed in Chapters 1 and 3. The concept that weight genuinely resembles is gating: the opening or closing of a generative channel, rather than the frequency with which an outcome occurs. Gating is discrete and can be hard-switched off; probability is continuous and normalized. The two must not be conflated. This section therefore establishes: probability belongs to the readout layer, as the statistics of many event outcomes, while weight belongs to the generation layer, as the generative reachability of a single relation; treating weight as probability is to mistranslate generative language into readout language.

6.3. The Unified Expression of Prehension, Coupling, and Coherence

The terms prehension, coupling, and coherence come respectively from process philosophy, dynamics, and quantum theory. Each is not incorrect in itself; the problem is that, in theory, they are usually layered too late—treated as three independent mechanisms rather than as three aspects of the same generative predicate. This section aims to show that, in generative ontology, these three can and should be unified into one generative predicate determined by differences.

Let us define them one by one. In this book, prehension does not take on its psychological meaning from process philosophy, but is strictly defined as generative receptivity: the degree to which one event, when updating, can receive the rhythm, phase, and direction of another event. Coupling is not an interaction term externally added between events, but the opening of a generative channel: the differences between two events fall within a compatibility interval, allowing the relation between them to enter generative settlement. Coherence is not taken in the geometric sense of quantum superposition, but is defined as rhythmic settleability: whether the rhythmic–phase differences among multiple events are stable enough to complete a consistent settlement. The three therefore reveal a common root: prehension is receptivity, coupling is the channel, coherence is settleability, and all are determined by the same set of differences—rhythmic difference, phase difference, and directional difference. They can therefore be expressed in a unified way and need not be treated as three independent mechanisms.

This unification is not a theoretical preference, but a necessity. Without unification, the three would be misread as three separate mechanisms that must be introduced independently, thereby violating conceptual economy (see the seventh criterion in Part Zero, "How to Falsify This Theory"); the unified expression reduces them to different readings of compatibility through differences. Correspondingly, their opposites are also unified: decoherence, decoupling, and refusal of prehension are likewise three sides of the same phenomenon—differences exceed the compatibility interval, the generative channel closes, and the relation exits settlement; they correspond uniformly to the weakening or clearing of weight. This section therefore establishes: relation is prehension, and prehension is a generative condition; the three traditional terms are unified into the same generative predicate, determined by the compatibility of differences between events.

6.4. The Network Is Not a Substitute for Space, but the Source of Space

Once an event network is introduced, a natural misunderstanding is to treat the network as a discrete substitute for space—as if continuous space had simply been replaced by a graph. This misunderstanding comes from still using spatial intuition to understand the network, and it must be corrected, because "substitute" and "source" are ontologically entirely different. "Substitute" means that network and space are two candidate backgrounds at the same layer; "source" means that the network comes first, while space is its frozen readout. This book adopts the latter: the network is not a substitute for space, but the source of space.

The key to understanding this is to see that, in the generative network, there is no pregiven "distance," only relational weight. Distance is a derived quantity that emerges from the statistics of weights at the readout layer after freezing, not an intrinsic property of the network. One might wish to use "graph distance" from graph theory, such as shortest-path length, as physical distance, but this does not work: graph distance is a quantity defined on an already given graph and presupposes the stability of the graph structure; by contrast, weights in the generative network are dynamically selected and can be cleared to zero. An unfrozen network has no stable graph distance. Therefore graph distance can serve only as a readout approximation after freezing, and cannot function as physical distance at the generation layer.

The true source of space is relational readout after freezing: when a large number of directional relations are frozen into stable structures, the readout statistics of these structures appear as "space." Space is the result of settlement, not the stage of organization. Dimension is likewise not an intrinsic property of the network, but a statistical quantity after a large number of directional relations have frozen. This is consistent with the conclusion of Section 5.3: direction precedes dimension, and dimension is the readout of frozen directional relations.

There is a key ontological asymmetry here, which most clearly shows that network and space belong to different layers: the generative network can locally collapse, with weights cleared and relations eliminated, whereas presupposed space in container grammar cannot disappear a priori. The network is a generative structure that can fail; space is the appearance after its successful freezing—one can fail, while the other is assumed not to. This asymmetry itself proves that the two do not belong to the same layer. This section therefore establishes: the network is not a substitute for space, but the source of space; space is the readout result of frozen relations, not the stage on which generation unfolds.

6.5. Continuity Is Not Ontology, but the Statistical Appearance of Successful Freezing

Chapter 2 has already suspended continuity at the diagnostic level. Now, on the basis of the event network and relational weights, this section can positively state the generative-layer source of continuity: it is not ontology, but the statistical appearance of many successful discrete generations after freezing, under finite-resolution readout. This section therefore complements Chapter 2: Chapter 2 explains why continuity cannot serve as a starting point; this section explains where the continuous appearance truly comes from.

Continuity appears indispensable because macroscopic readouts are almost always continuous—position, fields, and energy density all appear as smooth functions. But this continuity is the successful appearance at the readout layer, not an ontological premise at the generation layer. In generative ontology, continuity is not a condition of generation, but a result of generation: only when a large number of discrete generative events successfully close and freeze, and when their density is sufficiently high and their intervals sufficiently regular, does the readout layer present a continuous appearance. This appearance is produced by at least three mechanisms: high-density superposition of many discrete events, finite resolution of the readout protocol, which cannot distinguish discrete intervals, and the statistical regularity of frozen structures. Together, these three make discrete generation appear continuous at the readout layer. Once the continuous appearance forms at a certain scale, it becomes extremely difficult to disrupt, because the readout protocol itself operates at that scale and takes continuity as its default assumption. This explains why continuity is so persistent in physical practice—it is a stable appearance jointly locked by the readout protocol and frozen structures.

From this, the correct status of the continuous limit becomes clear: it is an effective approximation of frozen structures under a specific readout protocol, not the ontological starting point of generation. This is consistent with the conclusion of Chapter 2. It also reinterprets the failure of continuity: when continuity fails under certain conditions, such as so-called singularities, this is not a catastrophe in generative ontology, but an expected result—it only marks the point at which the frozen appearance no longer holds and discreteness at the generation layer reappears. This is consistent with the reinterpretation of singularities in Chapter 1. Finally, this section clarifies a common category mistake: dividing the world into "continuous" or "discrete" is itself mistaken—continuity is a readout appearance, discreteness is generative ontology, and the two belong to different layers rather than forming a same-layer alternative. This section therefore establishes: continuity is not an ontological property, but the statistical appearance of many successful discrete generations after freezing under finite-resolution readout.

6.6. What Relation Is: The Intrinsic Criterion of Generative Reachability

In the preceding sections, this chapter has systematically excluded all common misunderstandings: relation is not force, not probability, not spatial structure, and not a substitute for continuity. Now we must positively answer a question that has been deliberately postponed: in generative ontology, what exactly is relation, or weight?

This positive answer should be read directly from the Difference Principle in Section 5.5, rather than starting anew. Since the generation layer operates only through differences, and differences are quantities between events, "relation" is not a third thing outside events, but the form in which differences are organized and selected. The definition and update rule for weight given below are precisely the concrete form of this principle: all inputs read by weight—rhythmic difference, phase difference, and directional difference—are differences. Therefore, by definition, weight is relational, directed, and capable of failure. The task of this section is not to provide an intuitive explanation, but to provide a definition at the generation layer that is decidable, updateable, and capable of failure.

A fundamental point must first be clarified: in generative ontology, relation is not an "entity attached between two events." It is not a third object, nor an external attribute of events. Relation is meaningful only under a specific question: in the current generative state, can these two events still enter the same closure attempt? Therefore, relation is not "what it is," but an adjudication function. Accordingly, relational weight is defined in this book as the degree to which one event, when updating, can prehend the generative rhythm, phase, and direction of another event. More strictly, weight is not symmetric, not static, and not "relation strength"; it is a criterion of generative reachability. Compressed into one sentence: weight answers whether "this relation is still allowed to enter the next generative settlement."

Formally, weight must satisfy three minimal structural requirements: it is defined on the directed set of event pairs, its value represents the state of generative permission, and it is allowed to be weakened, frozen, or cleared to zero. Therefore the minimal form is introduced:

$$w : E \times E \longrightarrow [0,1],$$

where E is the set of events, $w(e_i, e_j)$ represents the generative prehension weight of event e_i toward event e_j during updating, and

$$w(e_i, e_j) \neq w(e_j, e_i),$$

so directionality is already contained here. This directionality is precisely the expression, at the level of weight, of "direction as the tendency of rhythm toward relational extension" in Section 5.3: the directionality of w comes from the asymmetry of directional difference, not from any external spatial orientation. It must again be pointed out why weight is not probability mathematically: although $w \in [0,1]$, it requires no normalization, no complete event set, no mutual exclusivity, and no repeatable sampling; the meaning of $w(e_i, e_j) = 0$ is that this relation no longer participates in any generation, not that it has an "extremely small probability."

Weight is not given, but is updated in the generative process. Its update does not depend on a time parameter, but on the compatibility of the triadic intrinsic structure, and—fully consistently with the Difference Principle of Section 5.5—all its inputs are differences:

$$w_{ij}^{(n+1)} = \mathscr{G}\left(w_{ij}^{(n)}, \Delta\phi_{ij}, \Delta f_{ij}, \vec{d}_{ij}\right),$$

where $\Delta\phi_{ij}$ is phase difference, the consistency of generative states; Δf_{ij} is frequency mismatch, the compatibility of rhythmic scale; $\vec{d}_{ij}$ is directional matching, the orientational bias of relation, namely the difference between the directional extension tendencies of two events; and $\mathscr{G}$ is the generative update operator, nonlinear and nonconservative. The key point is that every input term in the update rule is a difference between events; no absolute quantity of any event enters. This is the mathematical expression of "relation as operation on differences," and precisely because of this, the entire update rule can be rewritten in purely relational form without depending on any external reference (see the relation-only verification in Appendix B). Moreover, this update is not continuous evolution, requires no differentiability, and permits weight to drop abruptly to zero.

Weight therefore naturally connects to closure and freezing. In generative logic, closure does not "naturally occur," but is explicitly adjudicated. Its minimal condition can be stated as follows: there exists a candidate closure set $\mathscr{C} \subset E$, such that for all event pairs (i, j) in it, $w_{ij} \geq w_c$, where w_c is the generative threshold; if the condition is not satisfied, closure fails, and the relevant weights are weakened or cleared to zero, without introducing any probabilistic explanation. When a closure set repeatedly holds across multiple generations and its weights tend toward stable invariance ($w_{ij}^{(n)} \to w_{ij}^*$), this relation is said to be frozen; freezing does not mean that weight disappears, but that weight no longer updates, relation no longer participates in selection, and it enters the structure of the readout layer—force, space, distance, and continuity can appear only at this stage. The conclusion of this section is therefore definite: weight is not force, not probability, not a geometric quantity, and not statistical weight; weight is the intrinsic criterion of generative reachability, and its only responsibility is to determine which relations can continue to participate in generation, enter closure, and be frozen into structure. At this point, the proposition that "relations are not additions" has been fully closed at the ontological, logical, and mathematical levels—and its deepest basis is precisely the Difference Principle established in Section 5.5: relations are not additions, because the generation layer already processes only differences, and differences are essentially relations.

6.7. Summary of This Chapter

This chapter has completed a fundamental reconstruction of "relation" in generative ontology: through systematic exclusion and positive definition, it has established that relation is not a descriptive item added onto events, but an internal condition for whether generation can continue.

The chapter first rejected three historically deep-rooted misreadings. Relation is not force, because at the stage of generation there is no definable space, time, or motion; relation is not probability, because at the stage of generation there is no complete event set or statistical perspective; relation is also not space or a discrete substitute for space, because space acquires readout meaning only after freezing. On this basis, the chapter provided the ontological definition of relation, or weight: weight is not a pushing quantity and not a measure of uncertainty, but a criterion of generative reachability—it describes whether the prehensive capacity of one event, when updating, toward the rhythm, phase, and direction of another event is still allowed.

More fundamentally, this chapter connects the proposition that "relations are not additions" to the Difference Principle in Section 5.5: weight is jointly constrained by the differences in the triadic intrinsic structure of events—rhythmic difference, phase difference, and directional difference—and all inputs to its update rule are differences between events, not the absolute quantity of any event. Precisely for this reason, relation is not a bridge established between isolated events after they first exist, but the original form of operation at the generation layer: the generation layer processes only differences, and differences are essentially relations. From this, the update of weight is not continuous evolution, but generative selection; its clearing to zero is not a low-probability outcome, but generative elimination; its stability is not conservation, but completed freezing. Through this reconstruction, the chapter achieves several key unifications: prehension, coupling, and coherence are reduced to different aspects of the same generative predicate; the network is repositioned as the source of space rather than its substitute; and continuity is explained as the statistical appearance after successful freezing, not as an ontological premise. Ultimately, this chapter establishes an irreversible conclusion: in generative physics, relation itself is the condition of generation; events are not connected by relations, but are selected, closed, retained, or eliminated through relations. Relation is therefore no longer a language that describes the world, but the logical threshold through which the world becomes generated.

7. Cosmic Configuration and Logical Time

Events carry intrinsic structure (Chapter 5) and enter relation through differences (Chapter 6); but a single event, or even a set of relations, is still insufficient to constitute a "world." To rise from events to a world, one needs a unit that can be adjudicated and updated as a whole, as well as a way to discuss the advance of generation without presupposing time. This chapter provides that elevation: it first introduces cosmic configuration as the minimal unit from event to world (7.1), then establishes logical time as generative depth rather than a time parameter (7.2), and then rewrites, one by one, three concepts long occupied by temporal language—simultaneity (7.3), the arrow of time (7.4), and history (7.5). This chapter aims to show that all basic features of time can and must be reunderstood as results of generation and freezing, rather than introduced as presupposed backgrounds.

7.1. The Definition of Cosmic Configuration

The word "state" carries an inseparable structural burden in existing physics: in classical mechanics it depends on points in phase space, in field theory it depends on continuous functions, and in quantum theory it depends on vectors in Hilbert space. From the standpoint of this book, none of these structures can be introduced a priori. Therefore this book avoids the word "state" and uses "configuration" instead, in order to make three distinctions clear: a configuration is not coordinate assignment and does not assume background space; a configuration does not depend on continuous parameters or differentiable structures; and a configuration is a discrete unit that can be adjudicated and updated as a whole in the generative process. A configuration does not answer "what state the system is in," but only answers: what events and relations the current generative structure allows to hold simultaneously.

At logical generation step λ, the cosmic configuration is defined as a triadic structure

$$\mathscr{C}(\lambda) := (\mathscr{E}(\lambda), W(\lambda), \mathscr{F}(\lambda)),$$

where $\mathscr{E}(\lambda)$ is the set of events that can still participate in generation at that generative step, $W(\lambda) = [w_{ij}(\lambda)]$ is the structure of relational weights among events, and $\mathscr{F}(\lambda) \subseteq \mathscr{E}(\lambda)$ is the subset of events that have completed settlement and entered the frozen state. This definition deliberately excludes spatial coordinates, time parameters, probability distributions, and dynamical equations. A configuration does not describe "what the universe looks like," but only characterizes which events remain viable and which have been settled as facts at the current generative depth.

A misunderstanding must be clarified: a cosmic configuration is not a "snapshot" at a certain moment. In traditional intuition, a snapshot means that a background time exists, that the configuration is a capture of a certain instant, and that different snapshots are connected by continuous evolution. Here, all three points fail. The relation between configurations is not "change over time," but recursive production through generative rules:

$$\mathscr{C}(\lambda+1) = \mathscr{G}(\mathscr{C}(\lambda)),$$

where $\mathscr{G}$ is a set of discrete rules of generation–selection–freezing, not a differential or continuous evolution operator of any kind. The discreteness of configurations is not a modeling choice, but an inevitable result of generative logic: if configurations were continuous, continuous parameters would have to be introduced, and continuous parameters would immediately introduce limits and unfailing structures, which directly violate the basic requirements of generative ontology; only discrete configurations allow failure, bifurcation, and non-extendable termination, and failure is an indispensable component of generative logic.

The introduction of the frozen subset $\mathscr{F}(\lambda)$ is not a technical detail, but the core of the concept of configuration. Frozen events have three features: their intrinsic properties remain stable in subsequent configurations, they no longer participate in generative cascades, and they impose constraints on subsequent generation. Therefore, a configuration is not a "list of what currently exists," but a compressed expression of generative history at this step. The frozen subset gives the configuration structural memory without introducing any time axis. Precisely because time has not yet been defined, relations between configurations are not allowed to use the language of "before/after"; the only permitted relation is reachability: if there exists a finite number of generative steps such that $\mathscr{C}(\lambda')$ can be generated from $\mathscr{C}(\lambda)$ by the rules, the former is said to be reachable from the latter; the logical step λ is only an index for this reachable sequence, not a substitute for physical time. The level of configuration cannot be omitted—without it, events cannot be adjudicated as a whole, relations cannot be synchronously updated, and freezing cannot be recognized as structural fact; without configuration, there is no object called the "universe," only isolated generative attempts unrelated to one another. Configuration is therefore the minimal elevation from events to world. The core conclusion of this section can be summarized as follows: the universe is not an object evolving in time, but a generative sequence composed of configurations—time has not yet appeared, space has not yet appeared, but configuration already exists; configuration is not a way of describing the world, but the minimal structural unit through which the world can be progressively generated.

7.2. Logical Time Is Not a Parameter, but Generative Depth

The preceding section introduced the cosmic configuration $\mathscr{C}(\lambda)$ and pointed out that λ is only an index of the configuration sequence, not physical time. Yet this explanation alone is still insufficient to avoid misunderstanding: is λ merely a time parameter under another name? Does it still secretly play the role of an "evolution variable"? The aim of this section is precisely to remove this residual intuition and establish a strict conclusion: logical time is not a parameter, not a coordinate, and not an observable; it is only a count of generative depth.

In all theories based on differential equations, the time parameter t has three inseparable functions: it is continuous, it provides ordering for change, and it allows the definition of limits, instantaneous rates, and local evolution. But λ satisfies none of these: $\lambda \in \mathbb{N}$, so there is no continuous structure; the order of λ reflects only the steps of generation and does not correspond to "passing"; there is no limit $d\lambda \to 0$, and no "instant." Therefore, understanding λ as "discrete time" is still mistaken—"discrete time" is still time, while λ does not belong to the family of temporal concepts at all. Accordingly, logical time is strictly defined as the count of the minimal number of steps in which generative rules have been successfully applied from a given initial configuration. Formally, if $\mathscr{C}(\lambda+1) = \mathscr{G}(\mathscr{C}(\lambda))$, then λ records only the number of times that $\mathscr{G}$ has been applied and settlement has been completed. This counting definition deliberately avoids all physical semantics: it does not indicate duration, does not characterize rate, does not imply flow, and bears only one computational function—marking generative depth.

An attentive reader will immediately raise a sharp objection: when this book describes λ, it uses terms such as "step," "history," "before and after," and "advance," and these words themselves carry sequential semantics; if so, does λ, while verbally claiming to have no temporal semantics, secretly introduce temporal order through its language? This objection is justified and must be answered directly.

The key is to distinguish precisely which sequential intuitions λ retains and which it discards. λ is not completely unordered—it does carry a kind of order, but this order is the partial order of generative depth, not the total-order flow of time. It retains three sequential intuitions: first, incompressible cumulative depth, because each step of generation is a real advancement of settlement and cannot be skipped or merged; second, partial-order reachability, because if A freezes in $\mathscr{C}(\lambda)$, and B first becomes available in $\mathscr{C}(\lambda')$ with $\lambda' > \lambda$, then a generative influence of A on B is possible; third, directional asymmetry, which is established by the freezing monotonicity in Section 7.4. At the same time, it discards four sequential intuitions: uniform passage (there is no measure in which "each step has equal length," and λ has no unit), continuity and instantaneity (there is no $\mathrm{d}\lambda$, and no instantaneous rate), global simultaneity (the same λ in different systems does not mean "the same moment," and there is no unified present spanning the universe), and reversibility (the sequence cannot be inverted).

Therefore, in this book, terms such as "step" and "history" are strictly restricted to the meaning of cumulative depth in operations on differences, not to the meaning of temporal passage. The deeper basis for this is precisely the Difference Principle in Section 5.5: λ counts the number of times that operations on differences—encounter, comparison, update—have advanced; differences become operational quantities only when events encounter one another, and "encounter–comparison–update" itself constitutes the beat of generation. Therefore λ intrinsically belongs to the generation layer, not the ontological layer: the ontological layer is specification, and specification requires no steps; the generation layer is the advancement of operations on differences, and necessarily has steps. The kind of order retained by λ is precisely the partial-order depth accumulated by operations on differences; what it discards is precisely all the content of time as a continuous, uniform, global, and reversible parameter. When this book later uses the word "history," it always refers to this retained generative order constituted by already frozen acquired status, not to metric chronology. Logical time is therefore a count of generation, not a dimension of existence—it allows failure, because a step of generation may not succeed, whereas physical time never allows a "moment to fail."

7.3. Simultaneity Is Not a Slice, but a Settlement Condition

Once continuous time and the spacetime container are abandoned, a seemingly unavoidable question immediately arises: if there is no global time parameter and no slice of "now," what does "occurring simultaneously" mean? In existing theories, this question is often bypassed rather than directly resolved—either through coordinate conventions, as in relativity, through the definability of causal structure, as in Malament, or through measurement postulates, as in quantum mechanics. But all these paths share an implicit premise: some temporal structure already exists and only needs to be regulated or explained. In generative logic, this retreat is completely closed.

The concept of a "simultaneity slice" implies at least three premises: there exists a continuous or comparable time parameter, there exists a globally consistent standard of ordering, and there exists a set of already existing events that need only be "sliced out." In generative ontology, all three fail: there is no continuous time parameter, no global coordinate or external ordering, and events are not preexisting objects but results of generation. Therefore, "occurring at the same moment" is not a physical proposition awaiting explanation, but a sentence that lacks a domain of definition at the ontological layer. Simultaneity must be repositioned within generative structure. In this book, the minimal definition of simultaneity is: a group of events completes closure settlement within the same generative window. The core terms here are not "time," but generative window, closure condition, and completion of settlement. Simultaneity is no longer a coordinate relation, but the result of a generative adjudication.

This requires clarifying the meaning of "settlement." In generative logic, an event is not automatically established the moment it attempts generation. An event is counted into a configuration only when it satisfies three conditions: its intrinsic properties reach stability within a finite number of generative steps, its relational weights no longer undergo structural drift, and it no longer participates in an unclosed generative cascade. This state is called settlement. Settlement is not an "instant," but the fact that a condition has been satisfied; only in this sense is an event acknowledged as having "occurred." Accordingly, the criterion of simultaneity is a settlement window rather than a time label: given a group of events $\{\mathscr{E}_i\}$, if there exists a finite generative window $\Delta\lambda$ such that, for all i, the event $\mathscr{E}_i$ completes settlement within that window and their settlement result does not depend on the internal generative order inside the window, then this group of events is called simultaneous. It must be emphasized that "simultaneous" here does not mean "sharing the same λ," but "sharing the same closure window."

It follows that simultaneity does not require global consistency. A common misunderstanding is: if there is no global time, is a "global simultaneity" still needed to maintain a consistent world? The answer of generative logic is no—simultaneity is a local criterion; different event clusters can have different settlement windows, and there is no mechanism requiring the entire universe to share one "now." Simultaneity is therefore naturally multiple, local, and potentially inconsistent, while the consistency of the world does not come from "the same moment," but from the compatibility among frozen structures. This also marks the fundamental difference between this book and existing approaches: Einstein's simultaneity is an operational convention, defining coordinate time through the exchange of light signals; Malament's simultaneity is a structural problem, studying which relations can be uniquely characterized given a causal structure. The simultaneity in this book is neither a convention nor a theorem-object inside a given structure, but an earlier adjudicative fact—whether generation has completed settlement. It depends neither on light signals, nor on coordinates, nor on the a priori existence of causal structure. Logical time λ provides the generative stage for simultaneity, but does not define simultaneity itself: events within the same λ need not be simultaneous, events in different λ's may settle simultaneously, and simultaneity requires only that closure conditions hold, not that labels align. The traditional equation "simultaneity equals the same time label" is thereby completely dissolved. Finally, simultaneity is capable of failure: some events can never settle within the same window, some settlement windows are highly sensitive to generative parameters, and some structures that appear "almost simultaneous" in fact can never close. This capacity for failure makes "simultaneity" a genuine physical proposition rather than a semantic definition or coordinate choice. In short: simultaneity is not a temporal relation between events, but the fact that events jointly complete generation; it comes not from slicing, not from convention, but from closure.

7.4. The Arrow of Time Is Not Passage, but the Monotonicity of Freezing

Once simultaneity is rewritten as a closure condition, a more fundamental question emerges: if there is no continuous flow of time and no global clock, what does the so-called "arrow of time" point toward? In traditional physics, this question is usually outsourced to three types of mechanisms: the tension between microscopic reversibility and macroscopic entropy increase, as in statistical mechanics; special low-entropy initial conditions; and boundary assumptions at the cosmological scale. Generative ontology rejects all three paths, because they all presuppose an already established temporal structure before explaining "occurrence."

It must be stated in advance that the arrow of time is by no means an omitted issue in this book. It is forced by a non-negotiable condition—the irreversibility of freezing; see the third criterion in Part Zero, "How to Falsify This Theory." The argument has already appeared in dispersed form in Section 5.2.6, concerning the irreversible completion of phase settlement, and Section 5.3.7, concerning the fixed character of direction as a source of irreversibility, and it will be further developed in Chapter 13. The role of this section is to concentrate these dispersed arguments into a formal conclusion: the arrow of time is the monotonic retention of the frozen set.

First, "the passage of time" is not a physical concept. This expression implies at least three unauthorized premises: there exists a continuous parameter whose value changes monotonically; there exists a preexisting object that passively evolves along that parameter; and there exists an external or privileged perspective from which to mark "the present." In generative logic, there is no continuous parameter, no object that persists prior to generation, and no observer capable of standing outside generation. Therefore "passage" is neither a generative process nor a physical mechanism that can be defined, detected, or triggered. It is only a narrative restatement, at the readout layer, of a sequence of freezing.

The true source of irreversibility is freezing. Freezing is the entry of an event or event cluster into an absorbing state during generation, after which it no longer participates in generative cascades. Once freezing occurs, there is no generative rule that can restore it to an "ungenerated" state. This is not an empirical fact or a dynamical assumption, but a direct consequence at the definitional level of the concept of freezing: irreversibility is not introduced additionally, but is forced by the criterion of "completion and retention" itself. This irreversibility can be characterized more precisely so as to avoid the weak reading that "freezing is merely empirically difficult to reverse." The key structural distinction is this: what frozen structures prohibit is same-layer reconstruction—they no longer act as candidates, within the same settlement loop, bearing the adjudicative burden that they have already answered; but they still allow higher-level inheritance—as an already acquired whole unit, they can be incorporated into higher-level generative organization without reopening the internal freedom of adjudication. The irreversibility of freezing means precisely same-layer non-reopening, not "being unusable in every sense." An operational criterion follows: if a subsequent process reactivates any internal phase-update degree of freedom of a frozen unit while still claiming to treat it as already acquired, this is same-layer reopening, meaning that freezing has been broken; if it accesses that unit only as a coarse-grained whole without reactivating those degrees of freedom, this is higher-level inheritance, meaning that freezing is retained. It is precisely this distinction that makes the "monotonicity of the frozen set" below a precise and decidable structural proposition rather than a vague intuition.

Let $\mathscr{F}(\lambda)$ denote the set of events already frozen before logical generation step λ. From the definition of freezing, especially the same-layer non-reopening established above, it follows immediately that

$$\mathscr{F}(\lambda) \subseteq \mathscr{F}(\lambda+1),$$

that is, the frozen set is monotonically nondecreasing with the sequence of generative steps. This conclusion does not depend on any probability distribution, any coarse-graining or statistical averaging, any continuous limit, or any large-number assumption; it is a pure result of the generative criterion—more precisely, it is the structural consequence of the concept of "acquired status": what has been frozen remains in generative history as an acquired predecessor and does not return to the same-layer candidate role. Accordingly, the arrow of time in this book is no longer defined as entropy increase, energy dissipation, or an irreversible correction to microscopic dynamics, but as the monotonic growth of the number or complexity of frozen structures with the sequence of generative steps. The arrow of time is therefore no longer an externally added directionality, but the minimal and only necessary directional mark left by completed generation: generative history is no longer a neutral sequence of "earlier–later" stages, but an asymmetric structure of retention—later formation can inherit earlier frozen results as already acquired units, while earlier results do not depend on later formation to acquire their same-layer status. This non-reciprocity is sufficient to ground a minimal structural temporal asymmetry.

The relation between this arrow and the thermodynamic arrow must be made clear, because the two are easily conflated. A seemingly natural question is: is freezing monotonicity just entropy increase in another language? The answer is no. The key difference is that entropy is a statistical quantity that depends on state counting and assumptions of equiprobability; freezing is a structural fact that depends on definite generative criteria. Entropy increase can be reversed at the microscopic level; freezing is definitionally irreversible. Freezing requires no large-number limit and no "typical state"; it holds prior to any statistical description. More precisely, what this book establishes is not entropy increase, not the second law, not a theorem of statistical irreversibility, and not a metric time variable, but something earlier and more primitive—the retentive generative asymmetry of acquired status. Any later theory of temporal directionality, if it seeks to ground directionality rather than merely assume it, will have to respect or inherit this more basic asymmetry. This is the strength of this book's argument concerning the arrow of time: it does not compete with the thermodynamic arrow, but stands upstream from it. This also clarifies the relation between layers: freezing belongs to the generation layer, while entropy belongs to the readout layer. Only after frozen structures have already formed is there a "set of states" that can be counted, averaged, and assigned a distribution. Therefore, entropy increase is not the source of the arrow of time, but the statistical appearance of frozen structures after readout.

This arrow also has three properties worth noting. First, it requires no initial condition: traditional theories must appeal to a "low-entropy initial state," otherwise the arrow of time cannot be derived. Generative ontology requires no privilege of initial conditions—given any generative starting point, any initial event distribution, and any initial relational structure, as long as a freezing criterion exists, the monotonicity of the frozen set necessarily appears, and the arrow of time appears with it. Second, it allows local and multiple coexistence: freezing is a locally occurring generative fact, so different event clusters can have different freezing rates, and different regions can have different "time depths." There is no requirement that the entire universe obey a single synchronized arrow. This is consistent with Section 7.3, where simultaneity does not require global consistency—there is neither a global present nor a single global arrow. Third, it makes time reversal a pseudo-problem: in generative ontology, "time reversal" would mean returning from a frozen state to an unfrozen state, from a closure result to unclosed generation, from completed fact back to pure possibility. These operations are definitionally prohibited, namely by same-layer non-reopening above. Therefore, "time-reversal symmetry" is not a physical symmetry that needs to be broken, but an assumption that has never been legitimately introduced.

The conclusion of this section can be compressed into one sentence: time is not a parameter of passage, but the monotonic trace left after generation has been frozen. Finally, the strength and boundary of this argument must be stated clearly: what this book establishes here is a structural arrow of time—the minimal generative asymmetry produced by the monotonic retention of acquired status. It is not yet entropy increase, the second law, metric time, or the phenomenology of time. This limitation is not a weakness, but honesty: what the book claims is exactly the more basic asymmetry that later theories of time must respect if they seek to ground directionality, no more and no less. This also clarifies an external concern—the arrow of time is by no means omitted in this framework; it is forced by the core criterion of the irreversibility of freezing and has already been elevated into a formalizable monotonic law.

7.5. History Is Not a Record, but an Irreversible Generative Path

The preceding sections have systematically rewritten all the basic concepts of time: time is not a parameter, but generative depth; simultaneity is not a slice, but a settlement condition; the arrow of time is not passage, but the monotonicity of freezing. After this reconstruction, a question inevitably arises: if there is no continuous time axis, no global "now," and no reversible evolution, what exactly is "history"? In traditional physics and philosophy, history is almost always understood as a series of records of past states, a sequence of completed time slices, or an object "given as a whole" in a higher dimension; generative ontology holds that all these understandings are built upon premises that have been suspended.

Understanding history as "record" implies at least three assumptions: there exists a time axis that can be looked back upon, there exists some kind of preservable state snapshot, and there exists an external or meta-perspective from which these records can be read. In generative logic, all three fail: there is no time axis for retrospection, configurations are not state snapshots, and there is no omniscient perspective outside the generative process. Frozen events are not "recorded," but are no longer changeable by generation; history is therefore not what is preserved, but what is locked. Another common understanding—history as "the set of all events that have occurred"—also fails to hold: events do not gain independent existence simply by occurring; the meaning of an event depends on its position in the generative network, and the generative process does not produce "finished products" that can be listed side by side. The frozen subset $\mathscr{F}(\lambda)$ is not a warehouse of history, but the expression of irreversible constraints on generation. History is not equal to "the sum of frozen events," but to how those freezings constrain subsequent generation.

Accordingly, history obtains a minimal definition in generative ontology: history is an irreversible generative path; more precisely, it is a reachable chain in the configuration sequence constrained by freezing monotonicity. This chain has three core features: irreversibility—once freezing occurs, the generative path cannot return to the unfrozen state; incompressibility—generative steps cannot be skipped or reparametrized; and non-surveyability—there is no perspective that simultaneously contains all configurations. History is not an object, but the fact of how generation has once been restricted. This also makes "going back" a pseudo-concept in generative ontology: going back means defreezing, defreezing means denying the definition of freezing, and denying freezing means denying the irreversibility of generation. Therefore, "returning to the past" is not a capacity that has not yet been realized, but a statement that has no meaning in generative ontology.

This view of history breaks fundamentally and structurally—not merely interpretively—with the block-universe view. The block-universe view holds that all time points "exist simultaneously," that history is a worldline in four-dimensional spacetime, and that past, present, and future are merely differences of perspective. Generative ontology recognizes only stepwise generation and denies any structure that is "given as a whole": the block universe presupposes complete existence, whereas generative ontology denies that there is a "cosmic object containing all history." The universe is not a completed whole, but an unfinished generative process. It must be clarified in particular that generative ontology does not deny the existence of narrative, memory, or records, but these are phenomena of the readout layer and the cognitive layer. At the ontological layer, history is not a story being told, not stored information, and not an observed past, but this: which generative paths have been closed and therefore are no longer possible. History therefore has only one physical meaning in generative ontology—it constrains the future generative space: the more freezing occurs, the fewer generative paths remain available; the more stable the structure, the lower the variability. History is not a background, but the accumulation of constraints. In short, history is not the recorded past, but the fact that generation can no longer be undone; it does not exist on a time axis, does not wait to be reviewed, and does not allow itself to be rewritten—history is only how generation has once been locked, and thereby determines how the world can next occur.

7.6. Summary of This Chapter

This chapter has completed a generative rewriting of time. It has not discussed the properties of time against an already existing temporal background, but has shown that every basic feature of time is a result of generation and freezing, rather than a presupposed premise.

This chapter first replaced "state" with "configuration," reunderstanding the universe as a generative sequence composed of configurations, $\mathscr{C}(\lambda) = (\mathscr{E}(\lambda), W(\lambda), \mathscr{F}(\lambda))$. Before time and space appear, configuration already exists; it is the minimal elevation from event to world. The chapter then strictly established logical time λ as the count of generative depth rather than a time parameter. Facing the legitimate question of whether "λ secretly introduces temporal order," the chapter precisely distinguished the sequential intuitions retained by λ—incompressible cumulative depth, partial-order reachability, and directional asymmetry—from the sequential intuitions it discards—uniform passage, continuity and instantaneity, global simultaneity, and reversibility. It also traced the basis of this order back to the Difference Principle: λ counts the number of times operations on differences have advanced, and therefore intrinsically belongs to the generation layer. On this basis, the chapter rewrote three concepts occupied by temporal language: simultaneity is not a time slice, but the fact that a group of events completes settlement within the same closure window; it is local, multiple, and capable of failure. The arrow of time is not passage, but the monotonic retention of the frozen set, $\mathscr{F}(\lambda) \subseteq \mathscr{F}(\lambda+1)$; it is forced by the same-layer non-reopening of freezing and is a structural asymmetry prior to entropy and upstream from the thermodynamic arrow. History is not the recorded past, but an irreversible generative path, whose only physical meaning lies in constraining the future generative space.

Running through this chapter is a unified reversal: events do not occur in time; time emerges from the generation and freezing of events. Configuration, logical depth, settlement windows, freezing monotonicity, and generative paths together form a framework that can speak of "before and after," "simultaneity," "irreversibility," and "history" without requiring a temporal background. At this point, Part II has established the ontological starting point of generation (event), its intrinsic structure (rhythm and its dimensions), its mode of operation (differences and relations), and its organizational unit and temporal structure (configuration and logical time). The next part will turn to the dynamical core of all this: how phase evolves, how generation closes, and how completed structures are frozen into facts.

Part III.

Closure Dynamics: How Existence Appears

Introduction

In the preceding two parts, this book has completed a systematic work of removal. Part I removed the legitimacy of objects, spacetime, and continuous evolution as ontological starting points, and established the position of generative ontology centered on events, relations, rhythm, and freezing; Part II further removed the a priori status of time parameters, global simultaneity, and the arrow of evolution, showing that they can appear only as readout results of frozen structures. At this point, a crucial fact has been established: any dynamical description that takes continuous time as its background and state space as its stage no longer qualifies as a starting point at the generation layer.

Yet this conclusion simultaneously gives rise to an unavoidable tension. If there is no continuous time, no presupposed state, and no external parameter of evolution, then the following three questions can no longer be avoided or outsourced: how does generation advance? How is completion adjudicated? How can existence be maintained rather than dissipating instantly? Parts I and II have forcefully shown what one cannot begin from, but they have not yet positively answered whether, and how, one can move from the cleared starting point toward a structured world. Part III is introduced precisely within this tension.

The posture of this part must first be clarified: it does not attempt to "add back" any structure that has already been removed. It does not introduce continuous dynamics, reconstruct phase space, restore conservation laws as starting points, or move probability or statistical assumptions forward into the generation layer. On the contrary, the aim of this part is stricter: accepting the conclusions of the preceding two parts, it provides a minimal generation–closure–freezing process that makes the fact of "the appearance of existence" itself decidable. Accordingly, what this part introduces is not a "dynamical model" in the usual sense, but the restricted concept of "closure dynamics." Here, "dynamics" does not describe states changing over time, but describes four things: how phase is updated as the minimal intrinsic variable of generation, how relational weights form memory in generative history, how closure is adjudicated within finite generative steps rather than asymptotically approached, and how freezing transforms a completion from "capable of failure" into "already factual."

It must be emphasized that this part does not treat failure as an anomaly. On the contrary, failure, defreezing, rupture, and unreachable paths are explicitly incorporated into generative logic, and will be shown to be necessary conditions for the world to appear as a finite, sparse, and irreversible structure. A process capable only of producing success would immediately collapse into an undifferentiated, nonscarce, and therefore structureless world. Thus, Part III does not provide a formulaic system for "how to calculate the world," but answers a more basic question: without continuous evolution as support, what kind of generation can be recognized as existence, and why must most generation fail? Only after this question has been addressed directly can any discussion of laws, approximations, probability, and experimental readout cease to hang in the air.

This part therefore unfolds as a clear logical chain. Chapter 8 gives phase evolution as the minimal dynamics and establishes closure within it as the genuine criterion of settlement, that is, the threshold of existence; Chapter 9 explains how relational weights are updated in generation, how they form structural attractors, and how direction enters this update; Chapter 10 integrates phase, weight, and direction into a synchronized minimal process of generation–closure; Chapter 11 establishes three institutional principles for this process—conservation, boundedness, and monotonicity—where monotonicity is the only technical source of irreversibility; Chapter 12 elevates closure into a criterion of existence, explaining what exactly it is that closes; Chapter 13 then provides the full picture of freezing and irreversibility. Running through all these chapters is the same discipline established in Part II: no proposition is to be introduced unless it can be rejected—every criterion must be capable of failure.

8. Phase Evolution: The Minimal Dynamics of Generation

Part II established the ontological starting point of generation (events), the intrinsic structure (rhythm and its dimensions), the mode of operation (differences and relations), and the organizational unit and temporal structure (configuration and logical time). But all this remains static: it explains what the world is composed of and how it is organized, but has not yet explained how generation advances. This chapter provides the minimal dynamics of this advance—phase evolution. It aims to show how, without presupposing objects, spacetime, or conservation laws, phase is updated as the minimal intrinsic variable of generation (8.1), how the phases of multiple events are selected through coupling into combinations capable of common settlement (8.2), under what conditions a group of events becomes, from "many," "one" retainable generative fact (8.3, the core of this chapter), how completed structures are frozen into enduring existence (8.4), and why the vast majority of generative attempts must fail, with failure itself being the precondition of a sparse and reversible world (8.5).

8.1. Phase Is Not Motion

In almost all modern theories, "phase" is by default understood as some by-product of motion or evolution: in classical waves, phase advances with time; in quantum mechanics, phase rotates under the action of the Hamiltonian; in field theory, phase is a local state variable of a field in spacetime. These understandings differ formally, but ontologically they share the same premise—phase is a variable of change of some already existing object in some already existing time. Generative ontology rejects this premise.

If phase is understood as a state of motion, then at least three conditions must be presupposed: there exists an already generated and sustainable object; there exists a spacetime background bearing the motion of that object; and there exists a continuously advancing and parametrizable time variable. But in the generative framework of this book, none of these three premises has yet been established at the level of phase—objects have not yet been generated, spacetime has not yet appeared, and continuous time has not yet been defined. Therefore, any theory that interprets phase as a "state of motion" has already introduced the result of generation itself before phase has even been defined; this is not a difference in explanatory order, but a circular starting point.

In generative ontology, phase is redefined as the internal settlement state of an event at the current generative step. This definition has three irreplaceable features. First, phase is intrinsic: it does not depend on any external coordinate, position, or reference frame; it is not a quantity "relative to something," but the event's own generative state. Second, phase is discretely updated: it updates only between generative steps, namely logical time, and there is no process in which it "smoothly rotates in continuous time." Third, phase is generative: it does not describe a fact that has already occurred, but determines whether the event still has the possibility of entering closure. For this reason, the traditional expression $\theta(t) = \omega t + \theta_0$ is not merely "approximately invalid" in generative ontology, but ontologically unusable—it assumes that continuous real-valued time t already exists, that frequency is a derivative with respect to time, and that phase change is differentiable and interpolable. All these premises fail. The proper expression of phase is not a differential equation, but a discrete generative mapping

$$\theta(\lambda+1) = \mathscr{U}\big(\theta(\lambda), \text{relational structure}\big),$$

where λ is only the index of generative steps, not any physical time parameter.

This requires a strict distinction between two concepts: change describes the different values of an already existing object under some parameter, whereas generation describes whether a fact can be settled and acquire the qualification of existence. Phase belongs to the latter, not the former. The update of phase does not mean that "something has happened," but that an event is approaching, or failing to approach, a generative condition capable of closure and freezing. This also means that phase must logically precede force, energy, and trajectory: the order in traditional physics is force → motion → state → result, whereas generative ontology reverses it completely—phase structure comes first, phase relations determine whether closure is possible, and only after closure is completed do stable structures readable as "force," "energy," and "trajectory" appear. Force is not the cause of phase change, but a readout interpretation after phase closure. For the same reason, phase evolution at the minimal level presupposes no conservation law: before closure, there is no object that could be conserved. Conservation laws are not generative rules, but expressions of statistical stability of frozen structures in repeated generation. In the final analysis, the only physical responsibility of phase is to determine whether generation can still continue— phase alignment keeps a generative path open, phase divergence closes it, and phase locking triggers the condition of freezing. Phase is not a variable describing motion, but a gating variable of generative viability; it does not change with time, but updates with generative steps. Phase evolution thereby constitutes the minimal dynamics of whether generation occurs. Closure dynamics now obtains a dynamical starting point that depends on neither objects, spacetime, nor conservation laws, and all subsequent discussion of coupling, cascade, and closure can unfold only on this basis.

8.2. Phase Coupling: Not Interaction, but Generative Selection

In the existing language of physics, "coupling" is almost unavoidably understood as a kind of interaction: particles interact through force fields, modes exchange energy through coupling terms, and degrees of freedom influence one another through the Hamiltonian. These formulations differ technically, but ontologically they share the same premise—there exist already completed objects that exert influence upon one another within a given background. Generative ontology rejects this premise.

At the stage of generation, there are as yet no sustainable objects, no distinguishable subject and object, and no definable causal direction or transmission path. Therefore, to say that "event A acts on event B" is not an imprecise statement in generative ontology, but an ontologically meaningless one. In generative logic, the only question that can be posed is: can multiple unsettled events jointly enter the next step of generation? This is not a problem of interaction, but a problem of generative selection. Accordingly, phase coupling is defined in this book as whether the phase configuration of multiple events satisfies the same closure criterion. It does not indicate whether energy is exchanged among events, whether information is transmitted from one side to another, or whether influence is "applied" or "responded to." It answers only one question—whether there exists a generative step in which these events can settle together, whether there exists a configuration in which the many can become "one." This is an adjudication of viability, not a dynamical push.

Several key misreadings can therefore be clarified. First, coupling is not continuous strength, but discrete gating: traditional theories model coupling as a continuous parameter $g \in \mathbb{R}$, but this is misleading at the generation layer—generation either occurs or fails, closure is either satisfied or not satisfied, and there is no intermediate fact of "partial closure." A continuous coupling constant can appear only after freezing, as a readout approximation of a stable structure, and cannot serve as a generative condition itself. Second, phase alignment is not attraction, and phase mismatch is not repulsion. The more precise formulation is that phase alignment allows common settlement, while phase mismatch prohibits common settlement. There is no "pulling closer" or "pushing farther away" here, because at the stage of generation there is as yet no "near" or "far." What exists is only an adjudication: whether entry into the same generative result is permitted. Third, phase coupling precedes causal relation. In classical mechanics and relativity, causality is regarded as a fundamental structure, whereas the order in generative ontology is precisely the opposite—phase configuration determines whether co-occurrence is possible, and only after co-occurrence is frozen does a structure readable as causality form. Causality is the product of successful phase coupling, not its constraint.

Phase coupling also has an important property: it does not require a "propagation speed." In traditional physics, every interaction inevitably involves the problem of propagation—the speed-of-light limit, the paradox of action at a distance, and disputes about nonlocality. Phase coupling in generative ontology has none of these problems, because it is not a process "from here to there," but an adjudication of "whether settlement occurs together." At the stage of generation, there is as yet no "here" or "there," no path, distance, or delay, and therefore no problem of propagation speed. This point can be stated more precisely at the level of differences. The only input to the adjudication of phase coupling is the phase difference between events, $\Delta\theta_{ij} = \theta_i - \theta_j$, not the absolute phase θ_i of any event. This is fully consistent with the Difference Principle established in Section 5.5: what enters generative operations is never an absolute quantity, but a difference. Two direct consequences follow. First, because coupling reads only phase differences, a global translation of absolute phase changes no coupling adjudication. This is the dynamical expression of "there is no global generative clock and no universal unified phase" (Section 5.2.4), and it is also the basis on which the phase evolution mapping $\mathscr{U}$ can be rewritten entirely in pure phase-difference form without relying on any external reference (for the formal verification of this purely relational rewriting, see Appendix B). Second, the reason the problem of "propagation speed" does not arise lies here as well: difference is an adjudicative quantity of common settlement when events encounter one another; it does not involve any quantity being sent from one place to another. Without an absolute positional reference, there is nothing that needs to be propagated.

Finally, two separations must be made. First, the fundamental distinction between phase coupling and probability: probability describes the statistical frequency of many frozen results, whereas phase coupling describes whether a single generation can occur. Phase coupling is not probability weight, but the logical condition of generative viability; probability appears only after phase information has been truncated by a readout protocol. Second, failure of phase coupling is not "weak action," but termination of generation: the vast majority of generative attempts will fail. Phase mismatch means that co-occurrence cannot form, that the generative chain terminates here, and that no "unfinished object" or residual entity is left behind. This directly explains why the world is not filled with arbitrary possible structures, why stable existence is extremely rare, and why freezeable structures are highly constrained. In short: phase coupling is not a mechanism of action, but a rule of generative selection; coupling transmits no entity, requires no speed, distance, or causal cone; and coupling failure means generative termination, not "weak interaction."

8.3. The Closure Condition: When the "Many" Become "One"

The fundamental divide between generative ontology and all existing physical theories does not lie in using different variables, equations, or forms of expression, but in whether a criterion of "completion" is provided. Without a criterion of completion, so-called generation can only degenerate into a rhetoric of continuous evolution. This section answers only one question: under what conditions does a group of events cease to be merely "many" and become one retainable generative fact?

Before unfolding the specific criterion of closure, the concept of "closure" must first be elevated to its proper argumentative status. It is correct to define closure as "settleability," but that is not yet strong enough—it does not yet explain why the "completion" marked by closure is a genuine settlement, rather than some kind of "stopping" or "appearing to be complete" in a process. This book therefore sharpens this requirement into an independent burden.

This burden is the transition problem. Even if a theory has already satisfied the minimal conditions of generative ontology (the four conditions in Section 4.6) and has kept the functions of the generation layer and the readout layer from collapsing into each other (see Chapter 14), an independent problem remains unresolved: how does what has been achieved in generation become something inheritable by subsequent readout, without being constituted in return by that subsequent readout? This is the transition problem. It is not a rhetorical problem of linkage, but a structural burden that must be answered by any theory that simultaneously claims that "the generation layer genuinely achieves" and that "the readout layer does not collapse into it."

The transition problem can be removed only through a genuine criterion of settlement internal to generative organization itself. For a criterion to count as "genuine," it must satisfy three requirements at once. First, it must be non-circular: it must be stated only in terms of quantities belonging to the generation layer itself—rhythmic differences, phase differences, and directional differences—without appeal to any subsequent readout. Second, it must be substantively nontrivial: the distinction it marks between "already settled" and "still pending" must be a real and consequential difference, not a renaming. Third, it must be structurally grounded: it must be forced by the framework's own generative rules, not externally appended. The closure of this book is precisely such a criterion: it fixes the answer to "when multiple events become one retainable fact" in the purely generative-layer condition that phase differences converge within the tolerance determined by the minimal generative rhythm, and that this condition is continuously satisfied rather than satisfied only instantaneously. The definition beginning in 8.3.2 below should be understood in this sense as a genuine criterion of settlement, not merely as "reaching some state."

This is necessary because, without such a criterion, the transition problem will reappear in two characteristic distortions, and these two distortions correspond precisely to two important external concerns. The first is constitutive inversion: subsequent readout is explicitly or implicitly allowed to decide "which prior status counts as achieved," so readout no longer acts upon a settled predecessor prior to itself, but instead participates in constituting the very status it should only reveal. This is the precise form of the concern that "the readout layer merely renames and relocates the measurement problem." The answer of this book is: closure, as a genuine settlement criterion, establishes the settled predecessor prior to any readout. Readout acts only upon it and does not constitute it. Therefore, the measurement problem is dissolved rather than moved elsewhere. The full development of this answer belongs to Chapter 14, but its criterial basis is established here. The second is closed internality: settlement is claimed to be purely internal, but no account is given of how this settled status can have consequences for anything beyond itself. Settlement is thereby sealed inside itself and cannot be transmitted downstream. This book avoids this distortion by freezing (Section 8.4), which preserves the settled structure as an acquired unit inheritable by a higher layer (see the distinction between the same layer and higher layers in Section 7.4), thereby being both prior to readout and transmittable downstream.

There is a convergence here worth noting. The family of criteria consisting of closure–freezing–settlement was first introduced in Section 4.6 as one local implementation of the fourth minimal condition, namely an internal threshold structure. At the same time, it also removes the transition problem here, namely the bridging burden. Two independent lines of argument—the requirement of a threshold structure and the requirement of transition—point to the same family of criteria. This convergence strengthens, though it does not by itself prove, the weaker claim that closure–freezing–settlement tracks a real burden rather than terminological convenience. The full statement of the definition in 8.3.2 and of the seven diagnostic tests for criterion and consequence that it satisfies—non-circularity, substantive difference, structural grounding, settled predecessor, avoidance of constitutive inversion, avoidance of closed internality, and nonaccidental downstream bearing—is given in Appendix A.

Once this argumentative status has been established, the content of closure can be stated. First, "evolution" is insufficient to define existence. In continuous dynamics, "existence" is often treated by default as the evolution of a state through time, and once a solution exists it is regarded as part of reality. But evolution can extend indefinitely without providing a termination condition; it does not distinguish "not yet completed" from "already completed," and it cannot provide a criterion of failure. Without a condition of completion, all intermediate states are treated equivalently, "existence" loses operational meaning, and failure is wrongly interpreted as "not yet observed." Therefore, evolution is not the definition of existence. Accordingly, closure is strictly defined as follows: there exists a generative step such that the intrinsic phase configuration of a group of events satisfies a common settlement condition and can therefore be retained without further generation. This definition contains four indispensable elements: discrete generative step (closure occurs in a generative sequence, not under a continuous limit), joint adjudication of multiple events (closure is not a monadic property, but a condition on a set), phase condition (closure is determined by phase structure, more precisely by phase differences converging within the tolerance determined by the minimal generative rhythm, rather than by energy extrema or dynamical equilibrium), and retainability (the result of closure is qualified to continue existing). Closure is not "reaching a certain state," but passing an adjudication. According to what has been established above, this adjudication is precisely a genuine criterion of settlement—it is non-circular, substantively nontrivial, and structurally grounded, and therefore the "already settled" status it adjudicates holds prior to any readout.

Closure also has several structural properties that must be clarified. It must be a holistic condition: closure involves co-occurrence, co-occurrence involves relations, and relation is not a monadic property; therefore, the closure condition must be a criterion that holds for an event set as a whole. There is no stable existence that is "partially closed but not closed as a whole"; that is only an intermediate state that has not yet failed. Its minimal and irreducible form is a phase loop: at a certain generative step, the sum of phase increments of a group of events satisfies an integer quantization condition. This condition is basic because it depends neither on space, nor on energy measures, nor on any external reference frame—it is a purely intrinsic criterion, more precisely a purely difference-based criterion, involving only the closure of phase differences and no absolute phase (consistent with 8.2.6 and Section 5.5). If the loop fails, then no matter how "close" the system may be in any readout sense, generation must terminate. This also requires two separations: closure is not symmetry (symmetry describes structural properties, while closure describes whether generation has completed), and closure is not an extremum condition (an extremum is the result of continuous optimization, whereas closure is a discrete passage adjudication). In generative ontology, there is no "asymptotic closure," only closure or non-closure. Correspondingly, closure failure is not noise, error, or insufficient observation, but a definite generative result: it means that this event combination cannot be retained, will not enter freezing, and will leave no object or residual fact. This is the first point at which failure is established as an ontological property; henceforth, any structure incapable of failure is automatically excluded from physical explanation.

From this, the technical meaning of "completion" can be given: completion is not a rhetorical description, but a generative fact that has successfully closed and can be frozen. The relation among the three is strictly layered—closure is the necessary condition, freezing is the sufficient condition, and completion is the state in which both hold. In the terminology of settlement, closure gives the point of "answerability," threshold satisfaction gives the condition of "successful completion," and freezing gives the "status" of the completed result as an acquired predecessor; together, the three constitute settlement. Finally, all of this leads to a fundamental conclusion about order. The usual order in traditional physics is that events occur in time and causal relations organize events. The order in generative ontology is precisely the opposite: phase evolution proposes candidate structures, closure conditions select viable combinations, freezing forms stable facts, and time and causality appear as readout structures. Time is therefore not the background of closure, but the statistical readout of closure sequences. This order has a direct and important consequence: the settled status established by closure is prior to time and causality, and also prior to any readout or observation. This is precisely the criterial basis on which this book avoids constitutive inversion—the settled predecessor holds independently at the generation layer, and readout, including measurement, can only act upon it afterward; it cannot decide in reverse whether it counts as achieved. The reason facts in the early universe can hold without any observer lies exactly here (see Chapter 14; this also responds to the unacceptable consequence of "no observation, no fact" in Section 4.4). In short: existence is not a state of continuous evolution, but a generative event that has successfully passed the closure criterion; and this closure criterion, as upgraded in this section, is a genuine criterion of settlement—it is non-circular, substantively nontrivial, and structurally grounded, marking a settled predecessor prior to readout. The world is composed of facts that have passed adjudication; generative attempts that fail to pass leave no trace, and failure is as fundamental a component of the theory as success.

8.4. Freezing: Why Existence Can Persist

If closure answers "when existence appears," freezing answers a stricter and more fundamental question: why does existence not disappear immediately? In generative logic, appearance is not the same as persistence—the vast majority of closures are only momentary successes that then immediately fail. Without a further mechanism, the world would consist only of generative sparks flashing briefly and vanishing.

Closure by itself is insufficient to guarantee existence. Closure is only the fact of an adjudication: the conditions are satisfied at a certain generative step, phase forms a loop, and multiple events are temporarily settled as "one." But closure itself does not guarantee that the condition will still hold at the next generative step, that the phase structure can resist perturbation, or that subsequent generation will not break it apart again. Closure answers only "whether it appears," not "whether it remains." Without an additional criterion, all generative facts would degenerate into instantaneous events, and the concept of "existence" itself would lose physical meaning. Accordingly, freezing is strictly defined as follows: once closure is established, the event set enters a restricted generative state in which its key intrinsic variables no longer update freely in subsequent generation, but participate in new generation as constraints. This definition contains four points: freezing occurs after closure (it is not the starting point of generation, but the successor state after successful closure); freezing is not stasis, but restriction (it does not eliminate updating, but restricts degrees of freedom in updating); freezing introduces the status of constraint (the frozen structure no longer participates in generation as a candidate, but as a condition); freezing makes the structure repeatedly referable (only freezing can form reusable generative primitives). In short, freezing is not "stopping change," but ceasing to be regenerated as a candidate.

Freezing is precisely the origin of "objects." In generative logic, a direct and unavoidable conclusion can be given: without freezing, there are no objects; without objects, there is no persistence; without persistence, there is no physical world. Therefore, objects are not basic entities, but frozen generative results—particles, systems, fields, devices, and even "experimental apparatuses" are ontologically frozen event sets, not the starting points of generation. But it must be clarified that freezing is not absolute, but gated: it is not an eternal state, and is maintained only under specific generative conditions. Once the constraint is broken, it can be released; defreezing is not an anomaly, but a normal path of generation. Freezing is therefore not metaphysical invariance, but conditional persistence.

Its minimal and irreducible mechanism is phase locking: when the post-closure phase structure in subsequent generation satisfies stable relative phase, phase deviations cannot accumulate and amplify, and perturbations are absorbed by internal coupling, then the event set is no longer adjudicated as "unfinished," but is elevated into a stable structure. Freezing is therefore not an externally imposed constraint, but a self-stabilizing result of internal generative structure.

Freezing necessarily introduces the arrow of time. Once freezing occurs, the generative sequence immediately obtains irreversibility: before freezing, events are candidates; after freezing, events become constraints, and constraints cannot be withdrawn back into candidates. This is not statistical irreversibility, but logical irreversibility at the definitional level—the arrow of time is not a thermodynamic assumption, but a direct logical consequence of freezing (this is the same claim as the freezing monotonicity law in Section 7.4 stated in another place). Freezing also does not necessarily succeed: freezing failure means that phase locking cannot be maintained, perturbation exceeds internal absorptive capacity, and the structure falls back into the candidate set. Such failure is not noise or error, but the normal condition of generative dynamics—stable existence is rare precisely because the conditions of freezing are extremely strict. The stability of freezing also varies with scale: at small scales, freezing is brief and defreezing frequent; at intermediate scales, freezing is stable and forms the world of objects; at large scales, freezing appears as geometry, temporal order, and background structure. Scale is not pregiven, but the statistical appearance of freezing hierarchy at the readout layer. From this, a precise and minimal generative definition can be given: existence = an event set that has successfully closed and passed freezing gating. With closure but no freezing, the world only flickers; with freezing but no closure, freezing has nothing to trigger it; only when both hold does existence persist. Existence is not given, but a successful and retained generative result.

8.5. Failure, Defreezing, and Generative Elimination

If only successful closure and freezing are discussed, generative theory will inevitably slide into a hidden teleological illusion. The real universe is stable not because everything tends toward success, but because failure is systematically allowed, absorbed, and forgotten. Failure is not an exception, but the precondition for generative logic to hold.

In generative dynamics, one basic fact must be made explicit: the vast majority of generative attempts fail. The forms of failure include phase being unable to form a loop, the closure window being too high-order to stabilize, closure being momentarily established but unable to enter freezing, and freezing being briefly established but immediately disintegrating. If these failures are interpreted as "noise," "error," or "unobserved events," the theory would be forced to introduce remedial mechanisms and exceptional rules. In this book, failure is explicitly set as the baseline state of generative logic, not a deviation from it. A question that appears sharp and is in fact decisive follows: if failure is so frequent, why is the world not submerged by the "debris" of failure? The answer lies not in some mechanism of cleaning or recycling, but in a more basic fact—failure never enters the layer of existence. Failed generative attempts share common features: they do not complete closure and do not pass freezing; therefore, they do not obtain the qualification to be prehended by subsequent events. In other words, failure is not eliminated; it is never counted. This is the fundamental divide between generative ontology and evolutionary or statistical accounts.

Existence is not granted once and for all. Freezing is not permanent authorization. Defreezing occurs when phase locking is broken in continuous generation, key coupling weights fall below the freezing threshold, or the structure no longer satisfies the conditions of stable reference. Once defreezing occurs, the event set returns to the candidate pool, and its "objecthood" immediately fails; subsequent generation no longer treats it as a constraint. It must be emphasized that defreezing is not destruction, but the withdrawal of generative qualification. The freezing–defreezing mechanism naturally introduces a structure of elimination, but it is completely different from biological evolution, adaptive selection, or probabilistic screening: the criterion of generative elimination is simply whether low-order closure can form, whether phase locking can be maintained under perturbation, and whether the structure can stably participate as a constraint in subsequent generation. It involves no purpose, adaptation, optimization, or probability maximization. Elimination therefore does not select the "better," but retains the "sustainable." This also corrects an intuition that misunderstands generation as an omnipotent mechanism—"as long as something is possible, everything will occur." The position of this book is precisely the opposite: generation is highly constrained. Its constraint comes from the integrality and consistency of closure conditions, the stability requirement of freezing thresholds, and the mutual constraints among multiscale phase and coupling. This directly explains why arbitrary structures cannot exist, why arbitrary symmetries cannot be maintained, and why arbitrary initial configurations do not "move toward a world."

Failure has another decisive feature: it has no history. Only frozen structures enter generative memory, and only frozen results constitute "the past." Therefore, the past is not the sum of all attempts, but the structural projection of successful freezing—history is not recorded, but selected out (consistent with Section 7.5, where "history is an irreversible generative path"). This also explains why the generated world is sparse rather than saturated: if generation did not fail, the universe would rapidly saturate into structureless noise. The generated world is sparse because closure windows are discrete and finite, freezing conditions are extremely strict, and defreezing may occur at any time. Sparsity is not a defect, but a necessary condition of stable existence. In sum, failure is not an anomaly but the default result; failure leaves no trace because it never enters the layer of existence; freezing is not permanent authorization, and defreezing is withdrawal of qualification rather than destruction; generative elimination involves no purpose or optimization and retains only what can be sustained; history is the selected result of successful freezing rather than a record of all attempts. We can therefore state: failure, defreezing, and elimination are not supplementary mechanisms of generative theory, but foundational conditions necessary for it to hold.

8.6. Summary of This Chapter

Chapter 8 has completed the most crucial step in generative ontology: without presupposing objects, spacetime, or conservation laws, it has provided a complete dynamical answer to how existence appears and persists.

This chapter began from phase evolution as the minimal dynamics, establishing that phase is not motion and does not change with time, but is the internal settlement state of an event; it updates only with generative steps through the discrete mapping $\theta(\lambda+1) = \mathscr{U}(\theta(\lambda), \cdot)$ and serves as a gating variable of generative viability. On this basis, phase coupling was rewritten as generative selection rather than interaction—it transmits no entity and requires no propagation speed (its only input is phase difference, so a global translation of absolute phase changes no adjudication), and coupling failure means generative termination rather than "weak action." The core of this chapter is closure: it was elevated into a genuine settlement criterion in order to resolve the "transition problem"—how what has been achieved can become inheritable by readout without being constituted in reverse by readout. As a genuine settlement criterion, closure is non-circular, substantively nontrivial, and structurally grounded; the "already settled" status it marks holds prior to all readout, thereby allowing this book to avoid two distortions: constitutive inversion, through which the measurement problem is dissolved rather than moved elsewhere, responding to the concern that "the readout layer merely renames and relocates it"; and closed internality, avoided by the higher-level inheritability of freezing. This family of criteria simultaneously resolves the threshold requirement of the fourth minimal condition in Section 4.6 and the transition burden here. The convergence of these two independent routes strengthens the credibility that it tracks a real burden. Finally, freezing transforms successful closure into persistent existence (existence = closure × freezing) and necessarily introduces an irreversible arrow of time; while failure, defreezing, and elimination are established as the baseline of generative logic—the vast majority of attempts fail, and failures never enter the layer of existence. This is the basis for why the world is sparse, finite, and irreversible. At this point, the core logic of Part III has taken shape: existence is not given, but a generative result that has successfully passed a settlement criterion and been retained through freezing. The next chapter turns to the other side of this process: how relational weights are updated in generation and form structural attractors that allow certain structures to hold repeatedly.

9. Weight Updating and Structural Attractors

Chapter 8 provided the minimal dynamics of generation: how phase updates, how coupling selects, how closure is adjudicated, and how freezing retains. But relational weights have so far appeared as given quantities—Chapter 8 used threshold conditions such as $w_{ij} \geq w_c$, but did not explain how weights themselves are shaped by generative history. This chapter completes that link. It asks: why do weights grow, and why are they cleared to zero (9.1)? How do repeatedly successful relations form structural attractors (9.2)? What formal rules govern weight updating, and how does direction enter this update (9.3)? How does freezing elevate stability into fact (9.4)? What place do failure and dissipation occupy in the theory (9.5)? And how should "possibility" be rewritten after rejecting sample space (9.6)? Running through the entire chapter is a unified anti-teleological position: structures exist not because they are selected, but because they repeatedly have not failed in generation.

9.1. Reinforcement Is Not Selection

When explaining the appearance of stable structures, the most natural and also most dangerous intuition is "selection": stable structures exist because they are selected, preferred, or optimized in some kind of competition. This section aims to show that this selection narrative has no definition at the generation layer and must be replaced by a stricter concept—reinforcement as "remaining," not as "being selected."

"Selection" has no definition at the generation layer because it presupposes at least three things that do not exist at the stage of generation: a simultaneously available candidate set, a global evaluation function used to compare better and worse, and a preference-bearing agent that performs the selection. At the stage of generation, candidates are not laid out simultaneously on a table awaiting choice; they are attempted successively in generative history. There is no evaluation function standing above all relations, and no agent making preferences. Therefore, to say that "a structure is selected" is not an imprecise statement in generative ontology, but an undefined one. The true meaning of weight reinforcement is entirely different: when a relation repeatedly participates in successful closure and is frozen, its weight grows; but this growth is not a reward, but a record—a record that this relation repeatedly did not fail under perturbation and remains survivable. Reinforcement is therefore local and path-dependent: it depends only on the generative history of this relation itself, not on any global comparison. Different generative histories produce different reinforcement results, and there is no predetermined optimal endpoint waiting ahead.

It is in this sense that "attractor" must be reunderstood. In generative ontology, an attractor is not a presupposed target, does not correspond to any minimum of a potential function, and is not a final state toward which the system "tends." It is simply a relational substructure within the weight network that has repeatedly succeeded and whose weights therefore continue to grow. The reason an attractor appears at the readout layer as a "law"—as if the system obeyed a certain law and tended toward it—is that repeatedly successful relations form a stable and repeatable structure, while the readout layer misreads this statistical repetition as purposive tendency. This is fundamentally different from evolutionary theory and statistical physics: selection in evolutionary theory uses fitness as its evaluation criterion and a population as its arena of competition; equilibrium states in statistical physics take energy minimization or entropy maximization as their objective. Reinforcement in generative ontology has no fitness, no arena of competition, and no objective function to be optimized; it records only whether "this relation is still alive." In short: reinforcement means "remaining," not "being selected";

the growth of weight does not point to "better," but only to "still surviving."

9.2. Weight Networks and Structural Basins

Reinforcement records the survival of a single relation, but stable structures in the world are not single relations; they are organizations of relations. This section provides the form of that organization—structural basins—and emphasizes their ontological status: an attractor is not a terrain in space, but a history in relations.

First, the location of an attractor must be clarified: it is not in space, but in relations. The traditional picture of an attractor presupposes a phase space; an attractor is a region in that space, and the trajectory of the system falls into it. But in generative ontology, no phase space yet exists, and therefore an attractor cannot be a spatial region. The weight network is composed of directed weights among events. It is not a graph drawn in some space, but the current ledger of relational qualifications. In this network, a "structural basin" is defined as a relational substructure whose relations mutually reinforce one another and possess recovery capacity under perturbation: when an external perturbation attempts to weaken some of its weights, the repeated success of the remaining relations within the basin can pull it back. This recoverability does not arise because there is a force pulling the system toward a minimum, but because the relations inside the basin have repeatedly succeeded in generative history, and reclosure after perturbation still tends to follow these already reinforced paths. What determines the basin? The reinforcement history of the relations involved and the freezing threshold, not any external terrain.

The "depth" of an attractor is thereby redefined: it is no longer the depth of a potential well, but the historical length of freezing and the height of the defreezing threshold. The deeper a basin is, the longer it has been repeatedly confirmed through success and the harder it is to defreeze through perturbation; it does not mean that it lies lower on some energy surface. The coexistence of multiple attractors is therefore the normal case rather than the exception: different generative histories shape different basins, which are mutually independent and can coexist. The world does not tend toward a single global optimum. When a structural basin is sufficiently deep and stable, it appears at the readout layer as "physical property"—a "thing" with stable attributes that can be repeatedly identified. Physical property is therefore not an intrinsic attribute of an object, but a readout of the stability of a structural basin. The conclusion of this section can be compressed as follows: an attractor is not a terrain, but a history—it is not a pre-drawn valley into which the system falls, but a stable organization deposited in generative history by repeatedly successful relations.

9.3. Formal Rules of Weight Updating: From History to Constraint

If weight is not force, not probability, and not interaction strength, then an unavoidable question arises: how exactly is weight updated? Without explicit update rules, "weight network," "structural basin," and "attractor" would all degenerate into descriptive metaphors. This section provides the formal rules by which weight is shaped by generative history without introducing any external background.

First, it must be clarified that weight updating is not a process of "evolving with time." In generative ontology, updating means whether, after a generative attempt has ended, a relation is still allowed to participate in the next round of generation. Weight updating is therefore not continuous change, but a reevaluation of relational qualification. Let the event set be $\mathscr{E}$, and define the weight matrix $W(\lambda) = [w_{ij}(\lambda)]$, where $w_{ij}(\lambda) \geq 0$ indicates that the relation between events i and j still has the qualification to participate in generation at generative step λ. Weight does not require normalization. A zero weight means that the relation has been eliminated, not that its "probability is zero." Weight does not characterize strength, but reachability.

Updating follows a minimal principle: it depends only on the result of the previous round of generation and distinguishes only three cases—joint closure and successful freezing, attempted closure but failure, and non-participation in the current round of generation. The three cases correspond to three rules. Reinforcement rule: if, at generative step λ, events i, j jointly participate in closure, closure succeeds, and the result enters freezing, then

$$w_{ij}(\lambda+1) = \min\big(w_{ij}(\lambda) + \alpha,\ w_{\max}\big),$$

where $\alpha > 0$ is the minimal reinforcement step and $w_{\max}$ is the saturation value. Importantly, reinforcement is not a reward, but a record—that this relation remains survivable under perturbation. Decay rule: if events i, j attempt closure in this round but fail, then

$$w_{ij}(\lambda+1) = \max\big(w_{ij}(\lambda) - \beta,\ 0\big),$$

where $\beta > 0$ is the decay step. When $w_{ij}(\lambda+1) = 0$, this relation is permanently removed from the current generative network; no "probability transfer" occurs, and no compensation to other relations is required—this is generative elimination, not random dilution. Inertia rule: if a pair of events is not activated in this round, then $w_{ij}(\lambda+1) = w_{ij}(\lambda)$. This ensures that changes in weight are fully driven by generative history, with no "background evolution" and no hidden flow of time.

The reinforcement and decay rules answer the question of how closure results feed back into weight, but they presuppose a prior question: at the moment events encounter one another, what determines whether they are more likely to close or to fail in this instance? The answer is the compatibility of the triadic intrinsic structure. As established in Section 6.6, all inputs to weight updating are differences—rhythmic difference, phase difference, and directional difference. Their compatibility can be decomposed into an integration across three dimensions:

$$K_{ij} = k_f(\Delta f_{ij}) \cdot k_\theta(\Delta\theta_{ij}) \cdot k_d(\mathbf{d}_i, \mathbf{d}_j).$$

This product structure remains unchanged in mathematical form, consistent with the implementations in Appendices E and F, but its conceptual framework must be understood according to the triadic rewriting of Chapter 5: these three kernels are not the multiplication of three independent factors, but the integration of the three dimensions of the same rhythmic comparison—the scale dimension, the state dimension, and the relational-direction dimension. Together, they answer one question: to what degree can the rhythms of these two events form effective coupling? Reading them as "three independent factors" would repeat the parallel misreading corrected in Chapter 5. The correct reading is that they are simultaneous projections of rhythmic compatibility across three indivisible dimensions.

The directional kernel k_d especially requires the correct ontological reason here, because it is most easily misread. A natural but mistaken reading is to interpret k_d as the degree of alignment between two spatial vectors. But under the framework established in Section 5.3, direction is not spatial orientation, but the tendency of rhythm toward relational extension. Therefore, the correct reason for k_d is this: each of two events carries relational direction as the intrinsic extension tendency of rhythm. When they encounter one another, whether the tendencies of their rhythms toward relational extension resonate with one another determines the degree of directional compatibility of coupling. k_d measures the "resonance" or "compatibility" between two extension tendencies, not the geometric alignment of two objects in some external space. The mathematical form, such as $k_d \propto \frac{1+\mathbf{d}_i \cdot \mathbf{d}_j}{2}$, need not change; but the change in ontological reason gives real support to the claim of background independence here, rather than leaving it to hold accidentally because a dot product happened to be used.

This requires an explicit welding statement: k_d reads only directional difference, not absolute direction. Fully consistently with the Difference Principle in Section 5.5, what enters k_d is the relative orientation between two directional tendencies, namely directional difference, not the absolute direction of either event. The dot product $\mathbf{d}_i \cdot \mathbf{d}_j$ appearing in the formalization measures precisely this relative compatibility, not the projection of either side in an external coordinate system. Its direct consequence is that a global rotation of all directional coordinates changes no value of $\mathbf{d}_i \cdot \mathbf{d}_j$, and therefore changes no k_d and no weight adjudication. This statement is a necessary "welding point"—it ensures that when readers, including those who challenge the framework on grounds of "geometric circularity," see the dot product, they will not suspect that the directional kernel presupposes space once again: the dot product here is only a bookkeeping device for writing directional difference and can be globally rotated arbitrarily without any physical consequence. The formal verification of this invariance, together with the purely difference-based rewriting of closure and freezing criteria, is given in the relation-only verification in Appendix B; Appendices E and F, when using coordinates in numerical implementation, also state this on the same basis.

The directional kernel also naturally handles a previously unstated problem: where does propagation direction come from? The answer is that propagation direction need not be specified separately at the ontological layer; it emerges from the comparison of relational directions between two events at the generation layer. The key lies in the directedness of weight (Section 6.6): in general,

$$k_d(i \to j) \neq k_d(j \to i),$$

that is, the directional compatibility of one event toward another is not symmetric with the reverse. This asymmetry is not externally added, but is the natural result of the relative orientation of two extension tendencies in directed coupling. When this asymmetry accumulates and freezes within the generative network, it appears at the readout layer as "propagation direction"—the statistical appearance of the directional asymmetry intrinsic to relations. Therefore, propagation direction is the frozen readout of the asymmetry of directional differences at the generation layer, not the direction of motion in a presupposed space. This is consistent with Section 5.3, where "direction precedes dimension," and Section 6.4, where "the network is the source of space": direction, propagation, and even dimension are all readouts after relational-direction structures have frozen.

Weight updating must also work together with the mechanism of freezing. Suppose there is a freezing threshold w_{freeze} and a defreezing threshold w_{unfreeze}, with $w_{\text{unfreeze}} < w_{\text{freeze}}$: if $w_{ij} \geq w_{\text{freeze}}$, the corresponding relation can become part of a frozen structure; if $w_{ij} < w_{\text{unfreeze}}$, the freezing qualification is revoked, and the structure may defreeze. This hysteresis interval ensures that freezing is not an instantaneous and unstable label. It should be noted that "defreezing" here refers to the revocation of freezing qualification under the hysteresis interval, and should be distinguished from the discussions in Section 7.4 and the analysis of "same-layer non-reopening": an already acquired unit inherited by a higher layer is not reopened at the same layer because of weight hysteresis. The defreezing threshold in this section characterizes marginal relations that have not yet firmly entered acquired status—this point will be unified and clarified in Chapter 13.

Finally, the nature of these rules must be emphasized: they are not differential equations, not stochastic processes, and not optimization iterations. They do not describe "how the system evolves," but which relations obtain the qualification to continue existing within generative history—this is a logic of qualification, not dynamics. This rule set satisfies minimal completeness: it is computable (given generative history, it can be updated step by step), capable of failure (weights can be cleared to zero, structures can disappear), path-dependent (different histories produce different structures), and background-independent (all its inputs are differences, requiring no space, time, or probability). It is therefore the minimally complete formal rule for weight updating in GOCD. The conclusion of this section can be compressed into one sentence: weight updating is not an evolution equation, but the stepwise accounting of relational qualification by generative history—its compatibility is given by the integrated differences of the triadic intrinsic structure, its adjustment is driven by the success or failure of closure, and its change is driven entirely by generative history rather than by any background time.

9.4. Freezing, Irreversibility, and the Generation of Facts

Section 8.4 introduced freezing from the perspective of phase locking; this section completes it from the perspective of the weight network and establishes it as the stage of fact generation. The core claim is: stability is still insufficient to constitute existence. Only freezing elevates stability into fact, and the irrevocability of freezing is the minimal source of irreversibility and factuality.

Why is stability still insufficient? Because a structural basin may survive stably and succeed repeatedly, yet still remain in the status of a "candidate"—it may at any time fall back into the candidate pool because its weight drops below the defreezing threshold. Stability describes a dynamic capacity for survival, whereas fact requires a transformation of status. Freezing is precisely this transformation, but it is not deceleration or "stopping change" in the dynamical sense—the interior of a frozen structure may still update; what it ceases is "participating again in generation as a candidate." Therefore, the essence of freezing is role transformation rather than state change: before freezing, the structure is a candidate to be generated; after freezing, the structure becomes a constraint and begins to participate as a condition in the generation of others. This role transformation is the true source of irreversibility: a candidate can become a constraint, but a constraint cannot be withdrawn back into a candidate. This is not statistical difficulty of reversal, but logical irreversibility at the definitional level.

Accordingly, the arrow of time is not the direction of evolution, but the monotonic accumulation of the freezing sequence, consistent with Sections 7.4 and 8.4; it requires no entropy, no initial condition, and no statistical assumption. Freezing is therefore the minimal and sufficient criterion of existence: minimal, because no weaker condition can produce fact; sufficient, because once freezing occurs, fact is established. This also explains why failed structures leave no trace: what is not frozen never acquires the status of constraint and is therefore not prehended by any subsequent generation. It is not cleared away, but never counted. In this sense, freezing and "fact" are strictly identical: a fact is a frozen generative result, no more and no less. The true source of physical laws also lies here—so-called laws are the statistical regularities shown at the readout layer by many frozen structures of the same kind, not a priori laws governing generation. The conclusion of this section is: existence is not "being there," but "having been completed." It is not something standing still somewhere, but a generation that has successfully closed and been frozen, thereby becoming irrevocably a constraint.

9.5. Failure, Dissipation, and the Shadow Zone of Generation

If only successful reinforcement and freezing are discussed, the theory slides into hidden teleology. This section places failure, dissipation, and the "shadow zone" at the center of the theory, establishing an anti-teleological anchor: success and failure are completely neutral with respect to the generative rules, and the universe is not the collection of the successful.

Failure must be placed at the center because it is the default result of generation, not an anomaly. The vast majority of generative attempts cannot close, cannot freeze, and dissipate directly during generation. A failed structure can be strictly defined: it is a generative attempt that has not passed closure or has not passed freezing, and therefore has not obtained the status of constraint. Its decisive feature is that it does not accumulate and does not form a background—failure does not pile up like sediment to form some noise floor, because it never enters the layer of existence and has no status with which to accumulate. From this, the generative definition of dissipation can be given: dissipation is not the loss of energy, since energy itself is a readout after freezing, but the process in which a generative attempt fails to close and its phase organization is not retained but dissipates. Dissipation describes the occurrence of failure, not the transfer of some conserved quantity.

The ontological position of failure is called the "shadow zone" in this book: it is where all generative attempts that have not passed closure and freezing "are." But it must immediately be clarified that the shadow zone is not a container storing failures, but an ontological zone of negation. What it marks is precisely "not having obtained the status of existence." The shadow zone is in principle unreadable: a readout protocol can act only on frozen structures, and failed structures have never frozen, so no readout can touch them. The unreadability of the shadow zone is not a technical limitation, but an ontological consequence—readout has no object on which to act. This is precisely the central anchor of anti-teleology: success and failure are completely neutral with respect to the generative rules—the rules do not prefer success and do not avoid failure; they are simply applied, and then either closure occurs or it does not. There is no tendency that makes success "more deserving" of occurrence. The conclusion of this section is: the universe is not the collection of successful things, but is constituted by the traces that success has left. The vast majority of generation dissipates in the shadow zone; only a very small number of results that have not failed are frozen, counted, and read out.

9.6. Possibility, Unrealized States, and Generative Boundaries

The preceding section established that failure is the default result of generation and that successful freezing is only the rare trace that remains. This immediately raises a dangerous and unavoidable question: what exactly are the things that "could have happened but did not happen" in generative ontology? If this is handled improperly, the theory will flow back into sets of potential states, predefined sample spaces, or possible worlds that are "not yet realized but already exist." This section explicitly rejects these paths and provides a strict generative alternative.

"Possibility" is a high-risk concept because, in classical and quantum theories, it usually means an element in an already defined state space—unrealized, but already fully enumerated by the theory. Such understandings share an implicit premise: possibility already exists before the event occurs. Once this is accepted, one has accepted a space prior to generation, a complete set of possibilities, and a global perspective that is in principle accessible. These premises directly conflict with the generative position. Generative ontology therefore gives an assertion that must be stated explicitly: there are no "unrealized but already existing" states. Events that have not completed closure, structures that have not passed freezing, and generative attempts that have not been written into history have no ontological status. This is not an epistemic limitation, but an ontological adjudication concerning "what counts as existence." Therefore "not yet occurred" does not mean "already there": in generative ontology, "not yet occurred" means only that, in the current generative history, no structure satisfying closure and freezing conditions has yet appeared. It does not mean that some state is waiting in the background to be selected, or that some branch exists in potential form.

This leads to a key distinction: rules define generability, not an enumeration of existing states. Generative rules characterize generative boundaries—which phase–weight evolutions are legal, which closure conditions can in principle be triggered. But within these boundaries, there is no container filled with "possible states." A generative boundary is not a space, but a set of adjudicative conditions. This produces the deepest rupture with probability theory: what has not occurred has no right to be counted. In a probabilistic framework, an event that has not occurred still occupies probability mass, the sample space is fully defined before the experiment, and unrealized options still "count." In generative ontology, only structures that have successfully closed, passed freezing, and been written into history are qualified to enter any statistics or readout. Generative ontology therefore cannot accept sample space—the sample-space assumption holds that all possible outcomes have been listed before the experiment and that generation is only a kind of "sampling." In generative ontology, results are not sampled, but generated; and generation itself may fail. Therefore, there is no exhaustible set of outcomes or "complete in advance" space of possibilities.

Rejecting sample space does not mean giving up prediction, but prediction is strictly rewritten: it is not "what is the probability that result A occurs," but a constrained statement about generative rules and boundary conditions—which closures are impossible in principle, which thresholds a freezing must satisfy, and which readout structures must degenerate under changes of scale. The falsifiability of generative ontology comes from the failure of boundaries, not from statistical deviations. Correspondingly, the only legitimate meaning of "possible" is rewritten as "generable": under given rules and historical constraints, a certain class of closure is not in principle excluded. It carries no probability weight, no frequency meaning, and no commitment to existence; it is a purely generative and non-counting concept. The focus of theory thereby shifts fundamentally—from the problem of distribution to the problem of boundary: no longer asking "how are results distributed," but asking "which structures cannot freeze at all," focusing on impossible closures, unsustainable phase relations, and readout structures that must disintegrate under bandwidth changes. The conclusion of this section can be compressed as follows: in generative ontology, there is no possibility as an object; what is called "possible" is not an unrealized but already existing state, not a branch waiting to be selected, and not an option hidden in a sample space. It means only that, under the current generative rules and historical constraints, a certain class of closure has not yet been excluded. The universe does not walk through a space of possibilities; it continually attempts within generative boundaries and leaves behind only the very few results that have not failed.

9.7. Summary of This Chapter

The core task of this chapter has been to thoroughly remove an intuition that repeatedly returns in physics and philosophy: stable structures exist because they have been "selected." This chapter has provided a systematic alternative explanation.

First, the chapter rejected the legitimacy of the "selection narrative" at the generation layer: at the stage of generation, there is no simultaneously available candidate set, no global evaluation function, and no preference-bearing agent. So-called "reinforcement" is not a reward mechanism, but the record of compatibility repeatedly not failing in generative history—the growth of weight does not point to "better," but only to "still alive." Second, the chapter redefined the ontological status of attractors: an attractor is not located in any a priori space and does not correspond to the minimum of a potential function; it is a relational substructure in the weight network with recovery capacity under perturbation, and its "depth" comes from the historical length of freezing and the defreezing threshold, not from energy or probability. An attractor is not a target, but a trace of repeated success in history. Third, the chapter gave the minimally complete formal rules for weight updating—reinforcement, decay, inertia, and hysteresis—and supplied a key missing link: the compatibility that determines whether two events can close is given by the integrated differences of the triadic intrinsic structure, $K_{ij} = k_f \cdot k_\theta \cdot k_d$, where the directional kernel k_d is correctly understood as mutual resonance between two rhythmic extension tendencies rather than alignment of spatial vectors, and reads only directional difference, not absolute direction. Thus, propagation direction naturally emerges at the generation layer as the frozen readout of the asymmetry of directional differences, without presupposing space. This makes all inputs of weight updating, without exception, differences, fully connecting it with the Difference Principle of Section 5.5 and with Section 6.6, "relation as operation on differences." Fourth, the chapter clarified the relation between freezing and irreversibility: stability is not existence; only structures that pass freezing are written into generative history and acquire factual status. Freezing is role transformation rather than state change, and its irrevocability constitutes the minimal source of the arrow of time and factuality. Fifth, the chapter placed failure, dissipation, and the shadow zone at the core of the theory: failure is the default result; failed structures do not accumulate, leave no trace, enter no history, and form no background noise. The universe is not composed of the successful, but of traces left by success. Finally, the chapter rewrote "possibility" at the root: it rejects potential states, sample spaces, and possible worlds that are unrealized but already existing. What has not occurred has no right to be counted; rules define generative boundaries, not sets of states.

Chapter 9 thereby establishes a unified conclusion: structures exist not because they are selected, preferred, or optimized, but because they repeatedly have not failed in generation. This conclusion marks a decisive separation between this book and all theoretical frameworks based on potential states, probability distributions, and selection mechanisms. The next chapter will integrate the update rules of phase, weight, and direction into a synchronized minimal process of generation–closure.

10. The Minimal Dynamical Process of Generation–Closure

Chapters 8 and 9 respectively presented phase evolution, weight updating, and directional compatibility. Yet these do not occur separately in generation. This chapter integrates the three into an indivisible minimal process—the Generation–Closure Process (GOCD-Process). It aims to show that this process is not a list of rules, but an institutional closed-loop whole (10.1); within it, phase, weight, and directional extension constitute a synchronized three-process exploratory region (10.2); closure, as gating, is the sole threshold from "attempt" to "fact" (10.3); freezing enables facts to persist (10.4); and this process is precisely the minimal closure process that can neither be further decomposed nor requires anything more (10.5).

10.1. The Overall Semantics of the Generative Process

To understand the GOCD-Process, one must first reject an interpretation that reduces it to a "list of rules." Understanding the generative process as the sequential execution of five rules, D1 through D5, would immediately introduce an implicit order of execution and thereby an implicit time axis, directly contradicting the position of this book. Generation is not a linear procedure, but an institutional closed loop: the five rules are not steps executed successively, but mutually constraining aspects that hold jointly within the same generative step. Their relation is not "first do this and then do that," but "only when they hold simultaneously does this count as a generative step."

For precisely this reason, "partial execution" is illegitimate in generation. One cannot say, "first execute phase evolution separately, then execute weight updating separately," because the criterion of phase evolution itself depends on weights, while the basis of weight updating itself depends on the compatibility of phase and direction. If any one component is extracted and executed independently, it loses the source of its criterion within the closed loop and degenerates into an isolated rule requiring an external reference in order to operate. The minimal semantic unit of the generative process is therefore not any single rule, but one complete holding of the entire closed loop. This is also why the book insists on presenting it as a "process" rather than a "model": a model usually implies a set of equations or variables that can be tested separately, whereas a process emphasizes an indivisible whole—it either holds as a whole or does not hold. In short, generation is an institutional atom: logically, it cannot be further divided into independently operable components, just as an atom in the chemical sense cannot be divided into smaller units that still retain the properties of the element.

10.2. The Synchronized Three-Process Closed Loop of Phase–Weight–Direction

After establishing generation as an indivisible processual whole, one can still identify within this whole a minimal internally self-cycling structure. It does not directly generate existence, but determines whether generation remains qualified to continue. This structure is precisely the synchronized three-process closed loop of phase, weight, and directional extension.

It must first be explained why this is a three-process rather than a two-process structure, because this concerns the consistency between the generation layer and the ontological structure established in

Chapter 5. Expressing this internal closed loop as a two-dimensional "phase–weight" structure is insufficient: the intrinsic structure of an event is rhythm and its two dimensions—phase, as the internal state of rhythm, and direction, as the tendency of rhythm toward relational extension. If either is absent, rhythm cannot operate (Section 5.4). Correspondingly, the compatibility determining whether a generative attempt can continue must also be three-dimensional: rhythmic compatibility, phase compatibility, and directional compatibility must all be tested simultaneously. The compatibility kernel $K_{ij} = k_f \cdot k_\theta \cdot k_d$ in Section 9.3 expresses precisely this point. Therefore, the internal closed loop of a generative step cannot contain only the two aspects of phase and weight, but must incorporate the realization of directional extension as a third synchronized aspect. It must immediately be emphasized that this closed loop is not the "first half" of the generative process, nor is it a set of independently operable submechanisms; it is generation repeatedly testing its own viability before touching closure and freezing.

Phase, weight, and direction are inseparable from the outset. In language, we have discussed them separately: phase as the internal state variable of rhythm, weight as the memory of relational reachability, and direction as the tendency of rhythm toward relational extension. But in actual generation, the three never appear successively—there is no "phase first, then determine whether coupling occurs," no "weight first, then adjust phase," and no "direction first, then enter coupling." In the minimal generative step, phase, weight, and directional extension are always read simultaneously, tested simultaneously, and jointly determine whether the generative attempt continues. The reason follows directly from the unified structure of the triad: if phase is not expressed through weighted relations, there is no object of "joint generation"; if weight is not continually tested through phase evolution, it is merely a static marker; and if direction does not participate, the "capacity for alignment and qualification to continue" adjudicated by phase and weight remains orientationless alignment, unable to distinguish toward what relation an extension proceeds or whether it resonates with the extension tendency of the other event. Phase therefore indicates whether alignment is possible (the state dimension); weight indicates whether the attempt remains worth continuing (historical accounting); and direction indicates toward what relation the extension proceeds and whether the tendencies resonate (the relational-direction dimension). Only when the three are synchronized within the same closed loop do they jointly constitute a complete generative attempt.

The exact meaning of direction being "realized" within a generative step must be precisely defined, so as not to conflict with the conclusion in Section 5.3 that direction is fixed before freezing and is not updated by phase dynamics. In the synchronized three-process structure, the "realization of directional extension" does not mean that the directional value is updated at every step—direction, as the intrinsic extension tendency of rhythm, remains fixed before freezing. What is realized in the present step consists of two things in which this fixed tendency participates. First, directional compatibility is adjudicated: whether the relative orientation, namely the directional difference, among the fixed directional tendencies of events resonates is given by the directional kernel k_d, and thereby participates in weight adjudication. Second, relational extension actually occurs: the fixed directional tendency is realized in this step as an actual and oriented attempt at relational extension—the act of "extending toward this kind of relation" is genuinely attempted in the present step. The roles of phase and direction within a generative step are therefore asymmetric: phase updates at each step (dynamic), while direction is realized at each step but does not update (the repeated realization of a fixed tendency). This asymmetry is one structural source of why freezing is irreversible in Chapter 13—the stable structure locked by freezing relies on the fixed directional tendencies of its component events, rather than on continuous dynamical adjustment. The synchronized three-process structure is therefore not "three equivalent dynamical variables evolving in parallel," but "two dynamic dimensions—phase evolution and weight reassessment—and the realization of one fixed dimension—directional extension—cooperating within the same step."

A common misreading must be sharply excluded: this closed loop is not a dynamical process that "gradually tends toward stability." In traditional theories, a cycle often implies asymptotic convergence, a limiting state, or smooth evolution. In generative ontology, the semantics of this closed loop is entirely different—it describes whether repeated generative attempts are still permitted to continue. Each cycle of the closed loop answers only one question: under the current phase, weight, and directional configuration, is another generative attempt still warranted? Therefore, the closed loop may terminate abruptly, does not guarantee that it moves "increasingly close to closure," and in most cases fails directly within a finite number of cycles. This is precisely why generation is not evolution, but selection. Within this closed loop, a "generative attempt" must be strictly understood as a qualification test rather than a result: it does not mean that an object is forming, but only that, under current conditions, the event set remains permitted to participate in the next synchronized update and realization of phase–weight–direction. Failure of a generative attempt likewise does not mean that an "object has broken," because before closure there is no object at all. Failure means only that the combination has had its generative qualification revoked and no longer enters subsequent cycles. For precisely this reason, failure leaves no debris, does not accumulate, and does not enter history. The closed loop leaves only one result: whether continuation is permitted.

This closed loop must be internal: it depends on no external reference and requires no time parameter, spatial structure, energy function, or probability distribution. All its criteria arise from internal consistency and, consistently with the Difference Principle in Section 5.5, are expressed entirely in terms of differences—whether phase differences remain within the coherence window, whether weights remain above the generative threshold, whether directional differences still resonate, and whether history supports another attempt. Thus, this closed loop does not operate within some background; it is generation itself testing whether it can continue. This again confirms that background independence is mechanically realized: the closed loop accepts only differences, absolute quantities never enter, and therefore global translations and global rotations change no adjudication. A strict conclusion can thus be given: D1 (phase evolution), D2 (weight updating), and D3 (the realization of directional extension and the generative attempt) are logically inseparable. They are not three rules, modules, or stages, but three synchronized aspects of the same generative attempt, corresponding to the synchronized operation at the generation layer of the three dimensions of the triadic intrinsic structure. Any attempt to discuss one of them separately will unconsciously introduce external time or state space and thereby undermine the basic position of generative ontology; and any attempt to omit one of them, such as omitting direction, will render the generation layer inconsistent with the triadic ontological structure.

Finally, the fundamental limitation of this closed loop must be stated: by itself, it can never generate existence. No matter how many times the closed loop operates, so long as it has not touched the closure criterion, everything remains at the level of "not yet completed." This is precisely why closure is not part of the closed loop, but an institutional threshold—once triggered, generative logic immediately undergoes a transition. This section therefore concludes with one judgment: phase–weight–directional extension constitutes the exploratory region of generation, an internal synchronized three-process closed loop that repeatedly tests viability but has not yet produced existence. It simultaneously tests viability across three dimensions—state alignment (phase), relational qualification to continue (weight), and directional resonance (direction). Only when the next section introduces closure as a gating criterion does generation first confront the question of "whether it becomes fact."

10.3. Closure and Gating

The synchronized three-process closed loop continually tests viability, yet always remains at the level of "not yet completed." To cross from "attempt" into "fact," an element of an entirely different kind is required: closure as gating. This section aims to show that closure cannot be understood as the "natural result" of the closed loop, but must be understood as an institutional threshold whose triggering causes generative logic to undergo a transition.

Closure cannot be understood as a "natural result," because that phrase presupposes a continuous process of tendency—as though, if the closed loop ran for sufficiently long, closure would automatically appear as a limit. But this is precisely a residue of continuous grammar (Chapter 2). The necessity of gating lies in preventing generation from sliding into continuous evolution: without a discrete gate that is either passed or not passed, generation would degenerate into an asymptotic process of "approaching closure without limit." Such an asymptotic process never genuinely completes, and failure can always be deferred. Accordingly, the institutional definition of closure is: a discrete criterion that holds for an event set as a whole and, once satisfied, causes that event set to transition from "candidate" to "completion eligible for freezing"; if it is not satisfied, the attempt fails and exits the cycle.

Closure must be gating rather than a rule, because rules are repeatedly applied within the closed loop, such as the phase update rule at every step, whereas gating is a one-time adjudication of transition. It does not participate in the repeated exploration of the closed loop, but stands outside that exploration and determines whether the exploration has succeeded. When gating occurs, generative logic undergoes a transition—before it, the event set is objectless and remains a candidate that can be revoked at any time; after it, the event set becomes a completion eligible for freezing and prehension by subsequent generation. This transition cannot be partially satisfied: there is no "partial closure," just as there is no "partially passing through a gate." Closure either holds for the entire event set or does not hold; an intermediate condition is only "not yet failed." Closure and failure are therefore symmetric: at the same gate, passing yields closure and failing to pass yields failure. They are two results of the same adjudication. Failure is not the absence of closure, but a legitimate output of the closure adjudication. The conclusion of this section is: closure is the sole threshold between generation and existence—before it there are no facts; only after it can freezing and persistence be discussed.

10.4. Freezing and Persistence

Closure makes a group of events a "completion," but completion may still vanish immediately. This section aims to show that, even when closure succeeds, generation may still leave "nothing at all" without freezing. The introduction of freezing is not a patch, but a necessity; it is the sole mechanism through which existence can persist.

Why may nothing remain even after closure? Because closure adjudicates only that "this group of events has completed now," but does not guarantee that this completion will be retained in subsequent generation. If subsequent generation immediately breaks it back into candidates, then this closure is merely a momentary success that leaves nothing behind. The introduction of freezing is therefore not a patch intended to repair the theory, but a logical necessity: since closure provides only instantaneous completion, there must be a mechanism that transforms some completions into persistent constraints, otherwise "persistence" cannot arise. Accordingly, freezing is strictly defined as follows: a closed event set enters a restricted state in which its key intrinsic variables no longer update freely as candidates, but participate in subsequent generation as constraints. This is precisely a role transformation from "what is generated" to "a condition of generation." Before freezing, the structure is the object being generated; after freezing, it becomes a condition for generating other things.

It must be emphasized that freezing is not a dynamical process: it is not a kind of "deceleration" or "cessation of motion," because at the generation layer there is no motion at all. It is a transformation of status, not a continuous change of state. The true source of irreversibility in freezing lies here as well—a candidate can become a constraint, but a constraint cannot be withdrawn back into a candidate. This non-withdrawability exists at the definitional rather than statistical level (consistent with Sections 7.4, 8.4, and 9.4). Yet freezing is not eternal: it is gated, and once the weights maintaining it fall below the defreezing threshold, the freezing qualification may be revoked and the structure may return to the candidate pool. This gated character is not a defect, but is necessary for freezing to be a physical criterion capable of failure and therefore genuine. In the final analysis, freezing is the sole mechanism of "persistence": without it, the world is only a succession of unrelated momentary closures; with it, completions can accumulate, structures can form hierarchies, and history can arise. The conclusion of this section is: freezing is the institutional condition through which existence persists.

10.5. Why This Is the "Minimal Closure Process of Generation"

The final task of this chapter is to argue that the process constituted by the synchronized phase–weight–direction closed loop, closure gating, and freezing persistence is precisely the minimal closure process of generation—neither further decomposable nor requiring anything more.

"Minimal" has a precise meaning here: it does not mean "simplest" or "having the fewest steps," but "incapable of further subtraction." If any component is removed, the whole of generation–closure–persistence can no longer hold. This process cannot be further decomposed because the synchronized three-process closed loop, closure gating, and freezing persistence each perform an irreplaceable function: the closed loop explores viability, closure adjudicates viability as completion, and freezing retains completion as constraint. Without exploration, there is nothing to adjudicate; without closure, exploration never completes; without freezing, completion never persists. It is not an "evolutionary model," because it does not describe the continuous change of states with a parameter, but the discrete adjudication of qualification. Applying any "dynamical analogy" to it—such as understanding the closed loop as tending toward a limit or closure as a phase transition—is a category mistake, because every such analogy presupposes the continuous background suspended by generative ontology.

It cannot contain one fewer step: if direction is removed, the generation layer becomes inconsistent with the triadic ontological structure (Chapters 5 and 9); if closure is removed, generation never completes; if freezing is removed, completion never persists. The absence of each component corresponds to a failure already demonstrated in the preceding text. Nor does it require any further step: any proposed addition, such as “selection,” “optimization,” or “measurement,” either belongs to the readout layer rather than the generation layer or reintroduces a suspended background. The true meaning of the “minimal closure process” is therefore this: it is the minimal and complete generative structure required to make “how existence appears and persists” a decidable question capable of failure. The conclusion of this section is: in this sense, the GOCD-Process is irreplaceable—it is not one model among many possible models, but the minimally complete form that generation–closure–freezing can take under the constraints of the Difference Principle and the discipline of failure.

10.6. Summary of This Chapter

Chapter 10 has completed a crucial integration: it has combined the separately presented phase evolution, weight updating, and directional compatibility of the preceding two chapters into an indivisible minimal process of generation–closure.

The chapter first established that the generative process is not a list of rules, but an institutional closed-loop whole—the five rules are not steps executed successively, but mutually constraining aspects that jointly hold within the same generative step. Any “partial execution” would introduce an implicit time axis and violate the generative position. Within this whole, the chapter identified the minimal internal self-cycle: the synchronized three-process closed loop of phase–weight–directional extension. This constitutes the chapter’s crucial completion of the earlier draft—the compatibility determining whether a generative attempt can continue must be three-dimensional ($K_{ij} = k_f \cdot k_\theta \cdot k_d$); therefore, the internal closed loop of the generative step cannot contain only phase and weight, but must incorporate the realization of directional extension as a third synchronized aspect.

This "realization of direction" was precisely defined as compatibility adjudication and actual extension of a fixed directional tendency, not an update of the directional value. It is therefore consistent with the conclusion of Chapter 5 that "direction remains fixed before freezing," and it identifies the asymmetry in which phase updates at every step while direction is realized at every step but does not update as one structural source of the irreversibility of freezing. On this basis, the chapter established closure as gating—the sole threshold from "attempt" to "fact." It is a discrete, holistic transition adjudication that cannot be partially satisfied. Freezing was established as persistence—the sole mechanism that transforms momentary completion into persistent constraint. Finally, the chapter argued that this process is the minimal closure process of generation: it cannot be further decomposed, since removing any component causes the whole to fail, and requires nothing more, since any addition either belongs to the readout layer or reintroduces a background.

The dynamical core of Part III has thereby fully taken shape: generation repeatedly explores within the synchronized three-process closed loop, transitions into completion at the closure gate, and persists as fact through freezing. The next chapter will establish the institutional principles that this process must obey—conservation, boundedness, and monotonicity, of which monotonicity is the sole technical source of irreversibility.

11. Principles and Boundaries: Conservation, Monotonicity, and Irreversibility

Chapter 10 provided the minimal process of generation–closure. But for a process to constitute a self-consistent physical framework, rules specifying "how generation occurs" are not enough; it also requires a set of inviolable boundary conditions preventing rules from being abused, canceled, or circumvented. This chapter introduces three such principles: conservation (C1), boundedness (C2), and monotonicity (C3). They are not additional generative rules, but institutional constraints above the rules. This chapter aims to show why principles rather than more rules are required (11.1), what each of the three principles excludes (11.2–11.4), why monotonicity is the sole technical source of irreversibility (11.4), and why these principles are institutional permissions rather than provable theorems (11.5, 11.6).

11.1. Why Principles Rather Than More Rules Are Required

When confronted with a new phenomenon, the most natural response is to add a new rule to cover it. But this approach carries an institutional risk: rule proliferation. Each additional rule may expand the framework's coverage, but its unity, capacity for failure, and conceptual economy all decline. This is precisely what the seventh criterion in Part Zero, "How to Falsify This Theory," warns against. A framework that survives only by continually adding rules has already betrayed its own methodological standard. This book therefore introduces, beyond rules, a different kind of element—principles.

Rules and principles are fundamentally distinct. A rule is an operation repeatedly applied within a generative step, such as phase updating or weight updating. Rules directly participate in generation and can fail. Principles do not participate in any specific generation. They do not "do" anything, but specify "what kinds of generation are permitted." The institutional status of principles in this book is therefore analogous to that of a constitution relative to specific laws: a principle does not decide individual cases, but delineates the boundaries that no case may cross. It must be emphasized that principles are not "higher-order rules." A principle is not a metarule acting upon other rules, because a metarule remains a rule, still participates in some operation, and may still be canceled by another metarule. The special character of a principle lies in its not operating at the level of operation at all, but at the level of permission. Principles must be introduced here—after the generative process has been provided and before the criterion of existence is entered—because once the generative process is complete, it immediately confronts three questions that it cannot answer by itself: may generation proliferate without cost? May amplification become uncontrollable without constraint? May facts be rolled back? These questions cannot be answered internally by the generative rules, because they ask precisely "what constraints should rules be subject to?" Answering them is the function of the three principles of conservation, boundedness, and monotonicity.

11.2. Conservation (C1): The Generative Ledger

The first principle is conservation, but it must immediately be distinguished from the usual concept of conservation in physics: conservation (C1) is not conservation of energy, not an invariant, and not a Noether result of symmetry. These are all manifestations of frozen structures at the readout layer, whereas C1 precedes them. C1 means: every generation must be entered into a ledger; generation can neither occur from nothing nor disappear into nothing. This is a principle concerning a "generative ledger," not a law concerning some conserved quantity.

Why must generation be entered into a ledger? Because if generation could occur without cost—if arbitrary events, relations, and closures could appear from nothing without consuming anything—then generation would proliferate, the world would be indiscriminately filled, and sparsity (Sections 8.5 and 9.5) would be impossible. The minimal meaning of the generative ledger is precisely to assign an ineradicable record of cost to every generative attempt: generation consumes generative budget, closure occupies generative resources, and these expenditures are entered into the ledger and cannot be denied. It must be noted that conservation does not prevent generation, but prevents proliferation—it does not prohibit any specific generation from occurring, but prohibits generation from occurring indefinitely without paying a ledger cost. C1 therefore precedes every physical quantity: conserved quantities such as energy, momentum, and charge are concrete forms, at the readout layer, of what is entered into the ledger within particular frozen structures, whereas accounting itself exists prior to these forms. Finally, without a generative ledger there is no history: history is the accumulation of completed generation (Sections 7.5 and 9.4), and accumulation requires every completion to be entered and made ineradicable. If generation could occur and disappear without a ledger, nothing would be retained and history could not form. C1 is therefore the most fundamental condition making history possible.

11.3. Boundedness (C2): Preventing Systemic Degeneration

The second principle is boundedness. It answers a question that conservation alone cannot answer: even if every generation is entered into a ledger, the system will still degenerate if a single generation or structure can amplify without constraint. Without boundedness, conservation becomes ineffective—because conservation specifies only that "the total is accounted for," but does not specify that "an individual item may not grow without limit." An accounted quantity capable of infinite amplification would produce actual loss of control inside the ledger.

It must be clarified that boundedness is not a numerical upper limit, but an institutional constraint: it does not specify that a quantity may not exceed a particular number, but stipulates that no mechanism of unconstrained amplification may appear in the generative system. Boundedness prevents three paths of degeneration. The first is unlimited self-reinforcement of a single weight or structure: if one relation could reinforce itself without limit, it would monopolize all generative resources. The second is unlimited proliferation of a generative cascade: if one generation could trigger infinitely many subsequent generations without cost, the system would explode. The third is infinite subdivision or accumulation of scale: this would reintroduce the continuous grammar of "always being able to divide one step further." Boundedness must precede dynamical details, because it is a precondition for any dynamics to remain well-defined—a system allowing unconstrained amplification cannot possess stable dynamics at all. Boundedness is also related to the appearance of scale: precisely because amplification is constrained within a finite range, different freezing levels can stably separate and form structures readable as "scales" (Section 8.4). Finally, it must be clarified that boundedness does not mean that "the world is finite": it does not assert that the total number of events or the universe is finite, but only that amplification by any single mechanism is bounded. The world may be unbounded in the number of events, provided that no individual amplification becomes uncontrollable.

11.4. Monotonicity (C3): The Sole Technical Source of Irreversibility

Through the first two principles, this book has completed two exclusionary tasks: conservation (C1) excludes "costless generation," and boundedness (C2) excludes "unconstrained amplification." But these remain insufficient to explain one fact: even within a constrained and conservative generative system, why does history still possess directionality? This is the sole motivation for introducing monotonicity (C3).

Before developing this section, a map must first be provided to explain how the several appearances of the issue of "irreversibility" in this book divide their labor, because they can easily be misread as repetition. In fact, they are four different layers of the same conclusion. The first is the level of principle (this section), which establishes monotonicity as the non-negotiable principle C3 of the generative institution, namely that "rollback after freezing is illegitimate." The second is the formal level (Section 7.4), which formalizes this principle as the monotonic law of the frozen set, $\mathscr{F}(\lambda) \subseteq \mathscr{F}(\lambda+1)$, and uses the distinction between "same-layer non-reopening and higher-level inheritance" to make it a precise and decidable structural proposition, thereby deriving the arrow of time. The third is the mechanistic level (Sections 5.3.7 and 10.2), which identifies the structural basis of this irreversibility in the fixed character of directional tendencies—the asymmetry in which direction is not updated before freezing, while phase updates at every step and direction is realized but not updated at every step. The fourth is the level of full development (Chapter 13), which develops freezing as a complete mechanism. This section bears only the first level: establishing the principle. Its formalization and mechanism are not repeated here, but only cited where necessary.

Irreversibility cannot come from dynamics. In almost all existing theories, irreversibility is attributed to dynamical details: entropy increase arises from statistical coarse-graining, the arrow of time from a low-entropy initial condition, and breaking from symmetry selection. This book must explicitly reject this entire route for a simple reason: any irreversibility based on dynamics presupposes an already existing time parameter, whereas in this book time is itself one of the results requiring explanation. Therefore, irreversibility cannot be derived from equations of evolution; it can arise only from the generative institution itself. Accordingly, monotonicity is strictly defined as follows: once a generative fact completes freezing, it can exist only as a constraint in subsequent generation and cannot return to candidate status. In more technical language, the frozen set may only expand and never contract, the repository of facts may only increase and never roll back, and generative history may only accumulate and never be erased. This is a partial-order structure, not temporal evolution. It is precisely the type of order retained by λ in Section 7.2—the partial order of generative depth, not the total-order passage of time.

A key confusion must be clarified: monotonicity does not mean that "some quantity becomes larger." It describes the irreversible transformation of roles—before generation an event is a candidate, and after freezing it is a constraint; once the transformation "candidate $\to$ constraint" occurs, it cannot be revoked. Irreversibility lies not in numerical value, but in institutional status. In the language of acquired status, the frozen entity has transformed from candidate into an acquired predecessor and no longer returns to the same-layer candidate role. Without monotonicity there is no history: if a frozen structure could return to candidate status, today's facts could be revoked tomorrow, history would degenerate into a repeatedly editable draft, and "having already occurred" would lose technical meaning. In such a system, the arrow of time could not be defined, causal relations could not be fixed, and memory and structure would lose their foundation. The existence of history itself therefore depends on monotonicity.

Monotonicity must be strictly distinguished from thermodynamic irreversibility, because the two are easily conflated. The key distinction is that entropy is a statistical quantity dependent on state counting, whereas monotonicity is an institutional constraint dependent on freezing criteria. Entropy increase can be reversed at the microscopic level, whereas monotonicity cannot be reversed at the definitional level. In other words, entropy increase describes "how most paths are distributed," whereas monotonicity specifies "which operations are prohibited." This distinction receives a more precise position in Section 7.4: the monotonicity of this book is not a variant of entropy increase, but stands upstream from the thermodynamic arrow. It is "the most basic structural asymmetry that any later theory of temporal directionality must respect if it seeks to ground, rather than merely assume, directionality." Thus, this section (principle) and Section 7.4 (formalization) express the same claim at different levels: monotonicity precedes entropy, statistics, and any privilege of initial conditions. They are not repetitions, but two sides of "institutional permission" and "formal realization."

Monotonicity must be a principle rather than a rule. If monotonicity were written as a generative rule, it could be canceled by other rules, circumvented under certain configurations, or require additional conditions for maintenance. Irreversibility would thereby once again become a dynamical contingency. This book must therefore elevate monotonicity into an institutional principle: it does not compete with any specific rule, applies equally to every generative process, and stipulates only one thing—rollback is illegitimate. A strict conclusion can thus be given: in this book, irreversibility can arise only from monotonicity and not from any other structure—not from entropy, energy dissipation, time parameters, or initial conditions. Once freezing is monotonic, the arrow of time, causal direction, and historical structure all follow. It should be added that monotonicity as a principle stipulates that "rollback is illegitimate," but the reason it can hold rather than merely be declared has a mechanistic basis (Sections 5.3.7 and 10.2): the stable structure locked by freezing is retained through the fixed directional tendencies of its component events, rather than by continuous dynamical adjustment. Direction remains fixed before freezing and is not updated by phase dynamics; the asymmetry in which phase updates at each step while direction is realized at each step but does not update is precisely the structural source of why "the frozen cannot and need not be reopened at the same layer." The principle says that "rollback is illegitimate," while the mechanistic layer explains "why rollback has no structural route through which to occur." The two are complementary.

Monotonicity (C3) thereby completes the final institutional closure: C1 specifies that cost cannot be erased, C2 that scale cannot become uncontrollable, and C3 that facts cannot be withdrawn. The division of labor in the argument for irreversibility can now be clarified once and for all, eliminating the impression of "repetition across multiple chapters": Chapter 11 establishes monotonicity as a principle (institutional permission: rollback is illegitimate); Section 7.4 provides its formalization (the monotonic law of the frozen set plus same-layer non-reopening, deriving the arrow of time); Sections 5.3.7 and 10.2 provide its mechanistic basis (directional fixity and the asymmetry between phase and direction); Chapter 13 provides the full development of freezing. The four are not four restatements of the same content, but a progression of the same conclusion through the levels of principle–form–mechanism–development. At the same time, the boundary of the argument's strength must be stated clearly: what this book establishes here is structural irreversibility and temporal asymmetry—the minimal generative asymmetry produced by the monotonic retention of acquired status. It is not yet entropy increase, the second law, metric time, or the phenomenology of time. This limitation is not a weakness, but honesty: the book claims exactly "the more fundamental asymmetry that later theories of time must respect if they seek to ground directionality," no more and no less. Thus, the principle of monotonicity (C3) neither competes with thermodynamics nor pretends to have derived the entirety of temporal physics. It secures only one irreversible fact: at the most fundamental level of the generative institution, once a fact has been completed, it cannot be withdrawn.

11.5. Principles vs. Theorems: Why These Are Not "Provable Truths"

A natural objection is: on what basis does this book assert that the three principles of conservation, boundedness, and monotonicity hold? Have they been proven? The answer of this section is: they are not theorems, but principles; demanding that principles be "proven" is itself a category mistake.

Theorems and principles are fundamentally distinct. A theorem is a conclusion derived through logic within a given axiomatic system, and its truth depends on the truth of its premises. A principle is not derived within any system; it is precisely what defines "what kind of system is legitimate." In this book, principles are not descriptions—they do not describe how the world in fact behaves—but institutional permissions specifying what kinds of generation may enter this framework. Therefore, "proving a principle" is a category mistake. Demanding proof of the principle of conservation is like demanding proof of the first article of a constitution: the first article of a constitution is not a proposition proven within some higher legal system; it defines the legal system itself. Similarly, the three principles are not proven within some more fundamental generative framework; they define the category of a "generative framework" itself.

But "not provable" does not mean "not testable." The testability of a principle arises from boundary failure rather than successful verification: we cannot prove the conservation principle by "finding an example in which conservation holds," because no finite collection of examples constitutes proof; but we can falsify it by "finding a situation in which conservation is violated and the framework consequently collapses." This is the same logic as the failure criteria in Part Zero—the physical qualification of a principle lies precisely in its stating the circumstances under which it would be overturned. Precisely because principles are institutional permissions rather than empirical descriptions, they are not updated with experience: new experimental data can update laws and models within the framework, but cannot update the principles defining the framework. If some empirical finding forces us to abandon a principle, that does not "update" the principle; it abandons the entire framework and adopts another. The relation between principles and laws is therefore that "institution precedes content": principles do not tell us which specific laws the world obeys, but specify what any accepted law must satisfy—it must be accountable, bounded, and monotonic. The conclusion of this section is: these three principles are not "theorems that have not yet been proven," but something of a different kind—they are the institutional preconditions that make "theorems" and "laws" meaningful. Demanding proof of them is to mistake a constitution for a lemma awaiting demonstration.

11.6. Summary: The Institutional Boundaries of GOCD

This chapter has established three institutional principles for the generation–closure process and clarified their nature. Their roles can now be brought together clearly.

The role of the three principles is not explanation, but permission: they do not explain how any specific phenomenon occurs—that is the function of generative rules and freezing mechanisms—but specify the boundaries that every generation, every freezing, and every readout law must satisfy. For precisely this reason, they do not constitute a "system," but a boundary: conservation, boundedness, and monotonicity are not three propositions derivable from one another, nor an axiom set from which all physics can be deduced. They are three boundaries that must not be crossed, delineating "what kinds of generation can be admitted by this framework." This also clarifies the source of legitimacy of laws, models, and theories in this book: a law is legitimate not because it is confirmed by experiment—experiment confirms its effectiveness at the readout layer—but because its corresponding generative structure satisfies these three principles: it is accountable, bounded, and monotonic. A "law" that violates monotonicity, for example by permitting facts to roll back, is not accepted as legitimate within this framework, regardless of how well it may fit certain data.

This finally clarifies the position of this book: it is not another theory competing with existing physical theories, but is closer to an adjudicator. It does not provide new laws to replace old ones, but provides a set of institutional boundaries by which the legitimacy of any generative framework, including itself, can be adjudicated. The three principles of conservation, boundedness, and monotonicity, together with the generation–closure–freezing mechanism of the preceding chapters, jointly constitute the standards of this adjudication. This chapter has therefore completed the institutional layer of Part III: above "how generation occurs" (Chapters 8–10), it has established "what generation must obey." The next chapter will return to existence itself, elevate "closure" into the criterion of existence, and explain what exactly it is that closes.

12. Closure as the Criterion of Existence

The preceding chapters of Part III have repeatedly used the concept of "closure": it is the completion condition of phase evolution (Chapter 8), the basis of weight reinforcement (Chapter 9), and the gate of the generative process (Chapter 10). Yet what closure actually is, and what it adjudicates, still requires dedicated clarification. This chapter formally establishes closure as the criterion of existence. It aims to show that closure is not a cycle or return (12.1), that what it closes is the completeness of generation itself (12.2), that this completeness is adjudicable through the phase-closure condition (12.3), and that freezing transforms completeness from a single adjudication into a persistent fact (12.4). This chapter will also fix a set of terms used throughout the book: completeness, settled status, and settlement refer to the same criterion.

12.1. Closure Is Not a Cycle

The word "closure" readily evokes a geometric and cyclical image: a trajectory returns to its starting point, or a process loops back to its beginning. This image must first be removed, because it implicitly contains the container premise. The idea of a cycle presupposes a space in which a trajectory can be drawn, a position at which a "starting point" can be marked, and a metric by which "return" can be defined—and these are precisely the container structures that have been suspended. To understand closure as a cycle is to secretly reintroduce space into the concept of closure.

This book therefore makes a fundamental distinction: repetition is not closure. A periodically repeating process, such as a quantity repeatedly returning to the same value, does not thereby "close." Repetition is a property of a trajectory; closure is the satisfaction of a criterion. Closure is a criterion, not a form: it does not describe what the generative process looks like—whether it circles or is periodic—but adjudicates whether the generative process has completed. For precisely this reason, closure does not involve "returning to the starting point." It requires neither a starting point nor any return; it requires only that an internal condition be satisfied. This also explains why periodic motion cannot serve as a criterion of existence: a system in perpetual periodic motion completes nothing in any cycle; it merely repeats. If periodic motion were treated as the sign of existence, then "existence" would become equivalent to "continuous motion," which is precisely another version of "evolution is existence" under container grammar. The logical position of closure is therefore this: within the generative process, it adjudicates whether the process no longer needs to continue. It lies neither before the process, as an initial condition, nor outside the process, as an external ruling, but at the internal node where the process is about to stop. The conclusion of this section is: closure is not a form—a cycle, period, or return—but a criterion of whether generation has completed; understanding it as a cycle reintroduces the container from which it is intended to break free.

12.2. What Closure Closes: The Completeness of Generation

If closure is not a cycle, then an unavoidable question arises: what exactly does closure close? The answer of this book is both restrained and radical: closure closes neither an orbit, nor a variable, nor energy, but generation itself.

Before developing this section, a set of terms must first be fixed to prevent semantic drift across chapters. This book has used three apparently different but actually identical concepts: this chapter calls it completeness (*satisfaction / completeness*); Section 8.3 calls it the settled status adjudicated by a genuine settlement criterion; and in the criterion–bridging context it is called settlement. These three terms refer to the same thing: a generative process, under its own internal conditions, no longer requires additional generative steps to establish its identity and can therefore be acknowledged and prehended by subsequent generation without additional qualification. Throughout the main text, this book uses "completeness" as the standard expression for this concept, and "settlement" as its equivalent expression in the criterion–bridging context (Section 8.3 and Appendix A). The two are interchangeable and introduce no difference in meaning. Correspondingly, the three elements of completeness presented below are two formulations of the same set of conditions as the three requirements of the "genuine settlement criterion" in Section 8.3, and can be aligned one by one: the internality of this chapter—the criterion depends only on the structure of the generative process itself—is the non-circularity of Section 8.3—the criterion uses only quantities of the generation layer and does not appeal to subsequent readout; the termination of this chapter—generation no longer continues on the ground that it is "not yet complete"—is the substantive nontriviality of Section 8.3—the criterion marks a real difference between "settled" and "still pending," rather than merely renaming and postponing; and the recognizability of this chapter—the structure can be prehended by subsequent generation without additional qualification—is the structural grounding of Section 8.3—the criterion is forced by the generative rules themselves and can therefore bear downstream consequences without being sealed within itself. The reader may accordingly treat the criterion of "completeness" in this chapter and the criterion of "settlement" in Section 8.3 as two statements of the same criterion.

First, "completeness" is not an extremum condition. In existing physical theories, "holding," "stability," and "existence" are almost always adjudicated through some extremum condition: minimum energy, stationary action, minimum free energy, or maximum entropy. These conditions differ formally, but are logically highly consistent—they all assume an already given state space within which some scalar is compared. The "completeness" of this book does not belong to this category, for a simple reason: before generation has completed, the state space itself has not yet been defined; under such conditions, "smaller," "larger," and "better" have no domain of definition. Completeness therefore cannot be understood as "reaching the lowest energy," "the most stable configuration," or "the most symmetric state"—all these expressions presuppose a framework of comparison, whereas this book operates before that framework has been generated. Accordingly, completeness is strictly defined as follows: a generative process, under its internal conditions, no longer requires additional generative steps to establish its identity. This definition contains three indispensable elements, aligned one by one with the three requirements of the genuine settlement criterion in Section 8.3: internality—the criterion depends only on the structure of the generative process itself, namely non-circularity; termination—generation no longer advances on the ground that it is "not yet complete," namely substantive nontriviality; and recognizability—the structure can be prehended by subsequent generation without additional qualification, namely structural grounding. In other words, completeness is not "reaching a certain value," but the exhaustion of the reasons for generation.

Completeness is a criterion, not a property. A common misunderstanding is to treat completeness as a kind of "property"—a complete state, complete structure, or complete object. Such language is convenient but logically mistaken. Completeness is not an attribute "possessed" by an object, but the criterion by which a generative process is permitted to stop: once generation is adjudicated as complete, it is no longer regarded as "still being generated"; it obtains the qualification to be referenced, prehended, and constraining; and it enters the sequence of facts. Completeness is therefore not a label, but a threshold. This is fully consistent with the claims in Section 8.3 that "closure is the passing of an adjudication" and that "existence is a successful adjudication." Completeness is also not a "final moment." In many contexts, "completeness" can easily be misread as an endpoint, ultimate state, or final moment, but this book explicitly rejects that interpretation—completeness is not "the final step of generation," but "the completion of generation at this level." A structure that has already become complete may still participate as a substructure in higher-level generation, be reorganized at a larger scale, or disintegrate or decay in the future; but regardless of what subsequently occurs, this event of completeness itself cannot be revoked. The statement that "a complete structure may still participate in higher-level generation and be reorganized at a larger scale" must be understood strictly according to the distinction between "same-layer non-reopening and higher-level inheritance" (Sections 7.4 and 11.4): completeness prohibits the structure from being reopened as a candidate at the same layer—same-layer reconstruction is illegitimate—but permits it to be inherited and reorganized at a higher layer as an acquired unit—higher-level inheritance is legitimate. "Disintegration or decay in the future" refers to the reorganization of an acquired unit within higher-level restructuring, not to the reactivation of its same-layer degrees of adjudicative freedom, which is prohibited by the monotonicity of freezing. It is in precisely this sense that completeness constitutes an atomic source of the arrow of time: it marks a boundary of "having already occurred."

Completeness cannot be externally imposed. In traditional theories, whether a system is "complete" is often adjudicated by external conditions—boundary conditions, initial conditions, observational operations, or mathematical termination rules. This book permits no such external ruling. If a generative process requires an external declaration that "it may now be counted as complete," then it has not genuinely completed, but has merely been artificially truncated. Completeness must satisfy a strict requirement: if the external world did not exist at all, could this generative process still stop self-consistently? Only when the answer is "yes" does completeness hold. This requirement that the process can stop self-consistently even without an external world is the corresponding formulation in this chapter of the "settled predecessor prior to readout" established in Section 8.3. It is also the criterial basis on which this book avoids constitutive inversion: completeness is adjudicated by the generative process itself and is never constituted by subsequent readout, observation, or external ruling. Structures in the early universe can therefore become complete without any observer—their complete status is prior to and independent of all readout. The full development of this point appears in Chapter 14.

Completeness is also symmetric with failure. Its importance lies not only in permitting "existence," but also in making failure explicit. In this book, generation can fail; failure requires no additional explanation; and incomplete generation leaves no factual residue. Completeness is therefore not the default result, but a competitive result—the majority of generative attempts do not close, and only a very small number reach completeness under the constraints of phase, weight, and relation. It is at precisely this point that the book frees "existence" from necessity and makes it a scarce event. Finally, the relation between completeness and meaning must be noted. This book deliberately avoids discussion of value and meaning, but one point must be stated: completeness is not equivalent to "meaningfulness," yet every meaning must be grounded in completeness. An incomplete structure cannot be referred to, remembered, or compared and therefore cannot serve as the bearer of any structure of meaning. Completeness is thus the minimum threshold of factuality. This section establishes the second core conclusion of Chapter 12: closure closes generation rather than an orbit; completeness is not an extremum, stability, or symmetry, but the internal reason for generation to stop; without completeness there are no facts. It also fixes a terminological identity used throughout the book—completeness = settled status = settlement. The three refer to the same criterion and may be used interchangeably. Yet the question remains: if completeness is the reason for generation to stop, how is it adjudicated?

12.3. The Adjudicability of Completeness: The Phase-Closure Condition

If completeness is the reason for generation to stop, then it must satisfy a strict requirement: completeness cannot be merely a retrospective narrative, but must be adjudicated within the generative process. This means that the closure condition must be endogenous, computable, and non-semantic; otherwise, "completeness" would degenerate into an interpretive label rather than a criterion of existence.

Why should phase, rather than energy, bear the role of this criterion? In traditional physics, energy plays the central criterial role because it is a scalar, is conserved or changes monotonically in continuous time, and permits comparison through extrema. But in the generative logic of this book, all three conditions fail—without continuous time, energy flow cannot be defined; without presupposed space, energy density cannot be formulated; without a global state space, extremum comparison loses meaning. Phase is different: under background-free conditions, it possesses three irreplaceable features—intrinsicality, because phase belongs to the event itself rather than being an attribute embedded in space; comparability, because phase differences can be directly calculated within relational structure; and closure capacity, because phase permits strict and discrete periodic criteria. At the level prior to generation, phase is therefore the only natural bearer of closure.

The minimal formulation of phase closure is as follows: let $\{\mathscr{E}_i\}$ be a group of events undergoing generation, whose phases evolve with logical step λ. The minimal condition of phase closure is that there exists a finite step length $\Delta\lambda \in \mathbb{N}^+$ such that, for every event in the event set, its phase evolution satisfies an integer closure relation. In words: within a finite number of generative steps, phase returns to an identifiable identical relational state; no infinite approximation is required; and "approaching ever more closely without ever holding" is not permitted. This point is crucial because "finite-step closure" is irreplaceable. In continuous theories, closure is often weakened into the existence of a limit, average convergence, or asymptotic stability. All such conditions are unacceptable in generative logic for one reason: a limit can never fail. If "approaching without limit" were sufficient for completion, then any generation could be declared successful, and failure would lose its definition. This is precisely the failure-absorbing mechanism of continuous grammar diagnosed in Chapter 2. This book insists on a stricter but computable requirement: either closure occurs within finite steps, or it does not occur; there is no "partial completeness."

This phase-closure criterion also has several properties that deserve emphasis. It anchors generative identity: a group of events is "this particular" structure precisely because they jointly satisfy a specific phase-closure relation. Different closure relations constitute different structural identities; phase closure therefore adjudicates not only "whether completion has occurred," but also "what has completed." It makes failure explicit: because closure is a finite-step integer condition, "non-closure" is a clear and adjudicable result rather than an ambiguous state of "not yet converged." It is not periodic motion: phase returning within finite steps to an "identifiable identical relational state" means the reestablishment of a relational state, not a quantity periodically returning to its original value in space. The former is satisfaction of a criterion, while the latter is a form of trajectory, consistent with Section 12.1. It is also related to structural scale: phase-closure relations of different complexity correspond to multi-event structures of different scales. Simple closure yields small-scale structures, while higher-order joint closure yields larger-scale structures. Scale is therefore a readout of closure complexity, not a presupposed spatial hierarchy. The conclusion of this section is: completeness is a genuine criterion rather than a retrospective narrative precisely because it is endogenously adjudicable through finite-step integer phase closure. This criterion is computable, capable of failure, and independent of any external ruling.

12.4. Freezing: The Transition from Completeness to Fact

Completeness adjudicates that a generation has completed, but completion is still not equivalent to fact. This section aims to show that, even if completeness holds, existence cannot persist without freezing. Freezing is the transition from completeness to fact; it is not an artificially appended rule, but a necessary extension of the concept of completeness.

Why is completeness insufficient to guarantee existence? Because completeness is the fact of a single adjudication—it determines that "the reasons for generating this group of events have now been exhausted." But it does not itself guarantee that this completeness will be retained in subsequent generation. If subsequent generation immediately breaks this group of events back into candidates, then this completeness is merely a momentary completion that leaves nothing behind. The basic intuition of freezing answers precisely this problem: freezing means that generation ceases to participate in its own generation. Once a complete structure is frozen, it no longer participates as a candidate in generating itself, but participates as a constraint in generating other things. It must be clarified that freezing is neither an energy minimum nor a thermodynamic steady state. These concepts presuppose readout-layer structures such as energy, temperature, and ensembles, whereas freezing precedes them. Freezing is a transformation of status, not the attainment of an extremum by some thermodynamic quantity.

The minimal criterion of freezing is the repeatability of closure: a complete structure is frozen if and only if its phase-closure relation can be repeatedly confirmed in subsequent generative steps without disintegrating. This repeatability is precisely what distinguishes a "fact" from a "momentary completion"—a single completeness is an event; only completeness that can be repeatedly confirmed is a fact. Freezing thereby introduces the appearance of generative directionality: freezing is monotonic—the frozen does not return to candidate status, as established in Section 11.4—and the accumulation of the freezing sequence therefore produces an irreversible direction. This is consistent with the arguments in Sections 7.4 and 9.4. Freezing also provides the minimal definition of fact: a fact is a frozen complete structure—no more and no less. It must be emphasized that freezing is not an artificially added rule. It is not an additional assumption introduced so that the theory can "produce persistent objects," but a necessary extension of the concept of completeness along the dimension of persistence. Since completeness provides only instantaneous completion, "persistent existence" logically requires a stage that transforms completeness into something repeatedly confirmable; this stage is freezing. Finally, freezing is symmetric with failure: freezing can fail—a complete structure may be unable to be repeatedly confirmed and may disintegrate—and freezing can also be revoked through defreezing. It is precisely this capacity for failure that makes freezing a genuine physical criterion rather than a metaphysical device guaranteeing success. The conclusion of this section is: freezing transforms completeness from a single adjudication into a persistent fact; its criterion is the repeatability of closure; its monotonicity produces temporal direction; and its symmetry with failure guarantees its physical character.

12.5. Summary of This Chapter

Chapter 12 has completed a crucial conceptual transformation: it has detached "closure" from the traditional image of cycles and returns and made it a criterion of existence internal to the generative framework. In this book, closure is not a property of trajectories and does not presuppose a state space. It is an event of adjudication—whether the generative process has achieved a completeness that can be acknowledged without external supplementation.

The chapter first removed the cyclical image of closure: closure is not a cycle, period, or return, because all such concepts imply a container premise. Closure is a criterion rather than a form; it adjudicates whether generation has completed rather than describing what generation looks like. Second, the chapter answered what closure closes: closure closes the completeness of generation itself. The chapter fixed a terminological identity used throughout the book—completeness = settled status = settlement. The three refer to the same criterion, and the three elements of completeness—internality, termination, and recognizability—align one by one with the three requirements of the genuine settlement criterion in Section 8.3—non-circularity, substantive nontriviality, and structural grounding. Completeness is not an extremum condition, but the condition that "the reasons for generation have been exhausted." It is a criterion rather than a property, a threshold rather than a label. It cannot be externally imposed—this is precisely the criterial basis for avoiding constitutive inversion and allowing facts in the early universe to hold prior to observation—and it is symmetric with failure, thereby making existence a scarce event. Third, the chapter provided the adjudicability of completeness: phase, rather than energy, bears the closure criterion, and closure must be finite-step integer closure rather than asymptotic convergence, because a limit can never fail, whereas this book requires completeness to be capable of failure and computation. Finally, the chapter established freezing as the transition from completeness to fact: completeness is a single completion, while freezing—the repeated confirmation of closure—is a persistent fact. The monotonicity of freezing provides temporal direction, and its symmetry with failure guarantees its physical character.

The criterion of existence in Part III is thereby complete: existence is not given, not continuously evolving, and not confirmed by external observation, but is a generative completeness that has successfully passed the phase-closure criterion and has been repeatedly confirmed through freezing. The next chapter will fully develop freezing and irreversibility and fulfill the clarificatory commitments concerning the status of freezing made throughout the preceding chapters.

13. Freezing and Irreversibility

The concept of freezing has been repeatedly used throughout Part III, and several preceding chapters have repeatedly stated that certain questions would be "clarified uniformly in Chapter 13." This chapter fulfills that commitment and develops freezing in full. It aims to establish that freezing is not stopping, but a transformation in the mode of participation (13.1); that irreversibility arises not from time, but from freezing, while logical time provides only a rewritable partial order (13.2); that freezing introduces scarcity and ensures that the world is never saturated (13.3); and that failure constitutes the negative space of generation, shaping the boundary of success (13.4). This chapter concludes Part III by closing the logic of generation–closure–freezing into a complete picture of how facts become established and irrevocable.

13.1. Freezing Is Not Stopping

Freezing is one of the concepts in this book most susceptible to misunderstanding. The most common misunderstanding is to interpret freezing as "the stopping of generation"—as if, once a structure freezes, it exits generation and becomes an inert existent. This section establishes precisely the opposite: freezing is not stopping, but a transformation in the mode of participation.

Before proceeding, the commitments repeatedly embedded in several preceding chapters must be fulfilled. Sections 5.3.7, 8.5, 9.3, 10.2, and 12.2 all stated that certain questions would be "clarified uniformly in Chapter 13." These questions are in fact one set, centered on the status of a frozen structure and whether it can be reopened. This section provides a unified formulation once and for all, and all related statements in subsequent chapters should be understood accordingly.

In this book, a structure participates in generation in two fundamentally different ways. As a candidate, it is still testing whether it can close, and its internal phase, weight, and directional extension are repeatedly examined within the synchronized three-process closed loop (Section 10.2). As a constraint, it has completed closure and freezing, no longer tests itself, and participates in subsequent generation as an acquired condition. Freezing is precisely the transition from the former to the latter. A frozen structure therefore does not exit generation—it remains "within generation," but its identity has changed from candidate to constraint. Subsequent generation must prehend it and treat it as an acquired premise, while it no longer reenters its own generative degrees of freedom. The precise meaning of "freezing is not stopping" is therefore: freezing stops self-reflexive updating—the structure no longer updates its own internal phase degrees of freedom—but does not stop participation—the structure continues to participate in subsequent generation as a constraint.

This point must be expressed through the precise distinction introduced in Sections 7.4 and 11.4 and adopted as the unified formulation throughout the book. What freezing prohibits is the reopening of a frozen structure as a candidate at the same layer: it no longer bears, as a candidate within the same settlement loop, the adjudicative burden it has already answered—same-layer reconstruction is illegitimate. What freezing permits is its inheritance by a higher layer as an acquired whole unit and its incorporation into a higher-level generative organization without reactivating its internal degrees of adjudication—higher-level inheritance is legitimate. An operational criterion follows: if a subsequent process reactivates any internal phase-update degree of freedom of a frozen unit while still claiming to treat it as acquired, this constitutes same-layer reopening and freezing has been broken; if the process accesses it only as a coarse-grained whole unit without reactivating those degrees of freedom, it constitutes higher-level inheritance and freezing is preserved.

A frozen structure must therefore be strictly distinguished from a "state variable" in traditional theory. A state variable is defined on a presupposed state space, can take continuous values along a time parameter, and is in principle reversibly traceable. A frozen structure is not such a quantity: it is not defined on any presupposed state space—the state space itself is a readout of freezing, as shown in Section 6.4; it does not evolve continuously along any time parameter—it is an acquired result of discrete generation; and it is definitionally not reversibly traceable—it cannot be reopened at the same layer. To misread a frozen structure as a state variable is to downgrade acquired status at the generation layer into a reversible coordinate at the readout layer. This is precisely a residue of container grammar. More directly, irreversibility in this book is not the irreversibility of a state, but the irreversibility of a generative role: once a unit transforms from "capable of generating itself" into "generated only as a whole," there is no inverse operation. The structural reason that a frozen structure can be retained as a stable unit without continuous dynamical adjustment lies in the fixity of directional tendencies (Sections 5.3.7 and 10.2): direction, as the intrinsic extension tendency of rhythm, is fixed before freezing and is not updated by phase dynamics. The asymmetry in which phase updates at every step while direction is realized but not updated at every step allows a frozen structure achieved through stable directional organization to "maintain itself" rather than depend on external or continuous adjustment. Directional fixity is therefore the mechanistic source of why freezing neither needs nor permits same-layer reopening. It structurally supports the institutional declaration in Section 11.4 that "rollback is illegitimate."

Freezing is also the minimal starting point of "history" in this book. Before freezing, generative attempts leave no trace—failure does not accumulate, as established in Sections 8.5 and 9.5. Only freezing writes a completed generation into acquired status and gives technical meaning to "having already occurred." History is therefore not a record, but the accumulation of frozen results; it is not written, but selected out, consistent with Section 7.5. The term "defreezing," used previously in Sections 8.5 and 9.3, must now be clarified uniformly to avoid conflict with the irreversibility of freezing. "Defreezing" does not mean same-layer reopening, which is prohibited by the monotonicity of freezing. It has two legitimate meanings. First, defreezing under a hysteresis interval (Section 9.3) refers to the revocation of freezing qualification for a marginal relation that has not yet firmly entered acquired status—its weight lies between the freezing threshold and the defreezing threshold. Such a relation has never genuinely obtained acquired status, and its revocation therefore does not violate monotonicity. Second, defreezing in higher-level restructuring (Section 8.5) refers to the reorganization, as a whole unit, of an entity already inherited as an acquired unit at a higher level, as in decay or disintegration. This is a form of higher-level inheritance, not the reactivation of same-layer degrees of adjudication. Neither meaning constitutes a "settled predecessor returning to candidate status at the same layer," which is prohibited by definition in this book. The term "defreezing" is thereby rendered unambiguous throughout the book.

Finally, it must be explained that "freezing is not stopping" is not a matter of wording, but a logical necessity. If freezing were understood as "stopping," generative logic would immediately collapse for at least three reasons. First, the world could not remain stable: if freezing meant stopping, frozen structures would lose their connection with subsequent generation, and the world would fragment into unrelated static blocks. Second, hierarchical structure could not appear: higher-level structures depend on lower-level structures participating as wholes in generation; if freezing meant exit, hierarchy could not accumulate. Third, irreversibility would lose its source: stopping is symmetric—something can stop and restart—whereas irreversibility requires the one-way closure of a generative channel. A conclusion of decisive importance throughout the book can therefore be given: irreversibility arises not from time, entropy, probability, or observation, but from the generative operation of freezing itself. The logical chain is as follows: closure completes generation; freezing cuts off self-reflexive generation; the frozen structure can participate only forward in generation; generation thereby acquires directionality, without any additional assumption. Freezing also introduces another equally important irreversible bifurcation: successfully frozen generative units enter history, while unfrozen generation dissipates locally, and the two are no longer equivalent. Once a generation fails, it is not "remedied" in the future and does not return statistically; it simply leaves no trace. This explains a fact long overlooked: the world is not a superposition of all possibilities, but the residue after failure. This section therefore completes three crucial clarifications: freezing is not dynamical stopping, but a transformation of generative role; irreversibility arises from the one-way closure of a generative channel; and history and directionality first appear at freezing.

13.2. Irreversibility, the Arrow of Time, and Freezing: The Relation to Section 7.4

In existing physics, "irreversibility" is almost automatically equated with the arrow of time: time has directionality, and processes are therefore irreversible. The position of this book is precisely the opposite—the arrow of time is not the cause of irreversibility, but its representational result after freezing. The full argument for this position has already been given in the preceding text. This section provides only positioning and cross-reference, while adding one point not previously made explicit.

The argument that the arrow of time arises from freezing rather than from a time parameter is distributed across different levels of this book and has already been consolidated. Its formalization appears in Section 7.4: the frozen set is monotonically nondecreasing with generative steps, $\mathscr{F}(\lambda) \subseteq \mathscr{F}(\lambda+1)$, and same-layer non-reopening makes this a precise and adjudicable structural proposition, thereby deriving the arrow of time as the monotonic retention of acquired status. Its status as a principle appears in Section 11.4: monotonicity (C3), as an institutional principle, stipulates that "rollback is illegitimate." Its relation to entropy is likewise clarified in Sections 7.4 and 11.4: the structural arrow of this book stands upstream from the thermodynamic arrow; entropy is a secondary readout quantity of the sparsity of frozen structures, not the source of the arrow of time. These conclusions are not repeated here; the reader may refer directly to the corresponding sections.

The distinctive contribution of this section is to provide the converse of the argument in Section 7.4 and thereby highlight the irreplaceability of freezing. Section 7.4 derives monotonicity from the existence of freezing; this section considers the opposite case. Consider a world containing only generation and relational updating but permitting no freezing: every event always retains the capacity for self-reflexive updating, every relation can be rewritten, and no structure acquires acquired status. In such a system, any generative path can be erased by subsequent generation, no stage is "completed and no longer reenterable," and the ordering of logical time produces no persistent consequence. In this case, although there is "earlier/later"—the partial order provided by logical time, as discussed in Section 7.2—there is no irreversibility of any kind. This converse case establishes a key judgment: logical time itself provides only "ordering," not "irreversibility"; the entire source of irreversibility is freezing. The partial order of logical time is necessary scaffolding, but without freezing locking certain generative results into acquired status, this partial order remains rewritable and without persistent consequence. It is freezing, rather than logical time, that transforms "ordering" into "irreversibility" and rewritable generative order into monotonically accumulating history. This once again confirms the conclusions of Sections 7.4 and 11.4: freezing precedes time, and the arrow of time is the readout projection of the monotonicity of freezing. The conclusion of this section is: irreversibility arises not from a time parameter, but from freezing; logical time provides a rewritable partial order, while freezing locks it into monotonically accumulating history; and the arrow of time is the projection of this freezing monotonicity at the readout layer. The full formalization and principled argument appear in Sections 7.4 and 11.4 and are not repeated here.

13.3. Freezing, Scarcity, and the Unsaturation of the World

Freezing introduces not only irreversibility, but also another equally fundamental yet less frequently noticed consequence: scarcity. This section aims to show that freezing makes the world unsaturated—the world is never "complete," and this is not a defect, but a condition of existence.

Without freezing, the world would be saturated. Imagine a world in which all generative attempts are retained equally: every possible closure holds and leaves a trace. All possible structures would then coexist simultaneously, and the world would be indiscriminately filled, becoming a saturated whole without structural differences. Freezing breaks precisely this saturation because it introduces exclusivity: when one structure freezes as a constraint, it occupies certain channels of subsequent generation and excludes others. Its acquired status makes certain previously generable paths no longer reachable. Scarcity is therefore not insufficiency of quantity, but closure of paths. The world is sparse not because "there are too few things," but because every freezing closes some generative paths and increasingly constrains what structures remain reachable. This also explains why the world is never "complete": every freezing changes the topology of reachable paths, and generation always confronts an open frontier constrained by previous freezing. There is no terminal state in which "all structures have been generated."

Unsaturation is not a defect, but a condition of existence: a saturated world has no structure, because everything is present, and no history, because there is no selection. Only unsaturation makes it meaningful that "this structure exists while that one does not." Scarcity and irreversibility are internally isomorphic here: irreversibility means that "the frozen cannot return," while scarcity means that "the frozen has closed paths." The two are two sides of the same operation of freezing—freezing causes history to accumulate monotonically, which is irreversibility, and causes reachable paths to contract monotonically, which is scarcity. This also clarifies the relation to “conservation of energy” and “conservation of information”: these conservation laws are statistical regularities of frozen structures at the readout layer, whereas scarcity precedes them. Scarcity is not the finitude of some conserved quantity, but the more fundamental fact that generative paths are progressively closed through freezing. The conclusion of this section is: freezing introduces exclusivity and path closure, making the world sparse and unsaturated; unsaturation is not a defect of the world, but the condition under which it can possess structure and history; and scarcity and irreversibility are two aspects of the same consequence of freezing.

13.4. Failure, Rupture, and the Negative Space of Generation

Part III has repeatedly emphasized that failure is the default result of generation. This section elevates failure into a structural concept—the negative space of generation—and shows that failure is not noise, but a constraint geometry shaping the boundary of success.

What is the “negative space of generation”? It is the region corresponding to all generative attempts that failed to close or freeze and therefore “did not become facts.” But it must immediately be clarified, consistently with the shadow zone in Section 9.5, that negative space is not a container storing failures, but an ontological zone of negation. It is defined by “which generations were attempted but did not hold.” Failure is not an “unrealized possibility”: unrealized possibility presupposes a potential state waiting somewhere to be realized, precisely what Section 9.6 rejects. Generative failure means “attempted, not established, and leaving no trace”; it never existed as a potential state. Failure has three basic forms: closure failure, in which phase cannot form a finite-step loop; freezing failure, in which closure holds but cannot be repeatedly confirmed and disintegrates; and rupture, in which partially organized generation is interrupted by subsequent generation before completion.

The structural significance of failure lies in its shaping the boundary of success. This is the point of the metaphor of "negative space": just as the shape of a sculpture is defined by the parts carved away, the form of a successful structure is defined by the large number of failures surrounding it. A structure is "this particular" stable form precisely because countless neighboring variants have failed. Negative space is therefore not noise, but constraint geometry: it is not random background disturbance, but a structure that precisely determines "where success can and cannot occur." Negative space cannot be eliminated, because eliminating it would mean allowing all generation to succeed, which would immediately produce saturation (Section 13.3) and abolish all structure. This also gives meaning to the "thickness of history": history is not merely a sequence of successful freezings; its thickness also comes from the large number of failures it excludes. A history is "this particular history" rather than another precisely because countless unrealized bifurcations around it have been closed. The conclusion of this section is: failure constitutes the negative space of generation; it is an ontological zone of negation rather than a repository of potential states; failure appears in three forms, which jointly shape the boundary of success as a constraint geometry; and negative space cannot be eliminated, because it is precisely the condition under which structure and history take form.

13.5. Summary of This Chapter

Chapter 13 has completed the full characterization of the mechanism of freezing and fulfilled the clarificatory commitments concerning the status of freezing made throughout the preceding chapters.

The chapter first established that "freezing is not stopping": freezing stops self-reflexive updating, not participation; a frozen structure continues within generation in the role of a constraint. The chapter provided a unified formulation used throughout the book—freezing prohibits same-layer reconstruction, namely bearing again within the same settlement loop an adjudication already answered, while permitting higher-level inheritance, namely incorporation as an acquired whole unit into higher-level organization. On this basis, it clarified the two legitimate meanings of "defreezing"—the revocation of marginal relations under a hysteresis interval and the reorganization of acquired units within higher-level restructuring—neither of which violates monotonicity. Frozen structures were thereby strictly distinguished from "state variables": irreversibility is not the irreversibility of state, but the irreversibility of generative role, whose mechanistic basis lies in the fixity of directional tendencies. The chapter then clarified the relation between irreversibility and the arrow of time. The full argument was already given in Sections 7.4 and 11.4; this chapter supplied its converse—without freezing, logical time provides only a rewritable partial order and no irreversibility. The entire source of irreversibility is therefore freezing, and the arrow of time is the readout projection of freezing monotonicity. Finally, the chapter revealed two additional consequences of freezing: scarcity and negative space. Freezing introduces path closure, making the world sparse and unsaturated, while unsaturation is precisely the condition under which the world can possess structure and history. Failure constitutes the negative space of generation, shaping the boundary of success as a constraint geometry and remaining ineliminable.

Part III, Closure Dynamics, is now complete. Beginning from the minimal dynamics of phase evolution, it proceeded through generative selection by phase coupling, closure as a genuine settlement criterion, weight updating and structural attractors, the synchronized three-process minimal generative process, the three institutional principles of conservation–boundedness–monotonicity, and closure as the criterion of existence, before finally establishing freezing as the common source of factuality, irreversibility, scarcity, and history. A framework of generation–closure–freezing independent of container, continuity, and probability has thereby taken complete form: existence is a generative completeness that has successfully passed the settlement criterion, been repeatedly confirmed through freezing, and thereby become irrevocable. The next part will turn to a new question, one strictly constrained by this framework: how is such a generated and frozen world read out by finite readout—how do time, space, causality, and probability appear as structures of the readout layer?

Part IV.

The Readout Layer: How the World Is Read Out

Introduction

In the preceding three parts, this book completed two foundational tasks: it established generation–closure–freezing as the minimal dynamical semantics of physical existence, and showed that time, space, and causality are not a priori stages, but organizational results of generative structures at the readout layer. From this, the task of Part IV becomes unavoidable: once generation has completed and freezing has been established, why does the world appear as a statistical, probabilized whole with the appearance of "interaction"?

This question cannot be answered by introducing new entities at the generation layer, nor can it be avoided by adding probability, force, or randomness at the ontological layer. It can only be addressed through a strict analysis of the generation–readout interface. This is precisely the function of the concept of the "readout layer." The nature of the readout layer must be clarified from the outset in order to rule out a deeply rooted misunderstanding: the readout layer is not an observer, not consciousness, and does not presuppose any macroscopic measuring device. It refers to the truncating contact of any finite system with frozen structures under finite bandwidth, finite time window, and finite resolution. Once a generative fact is prehended by another generative structure in a restricted way, that is already a "readout." The readout layer is therefore everywhere in the universe and operates prior to any observer—this point is decisive for responding to concerns about the "pre-observer universe" (see Chapter 14).

This part contains a structural claim that must be clarified in advance and that bears on the success or failure of the entire book: introducing the readout layer does not rename and relocate difficulties that the generation layer failed to solve—especially the measurement problem—to a new layer; rather, it dissolves them already at the generation layer. This distinction—dissolution rather than relocation—is the pivot of Part IV. Its validity depends on the criterial basis already established in Sections 8.3 and 12.2, namely closure as a genuine settlement criterion, which makes the settled predecessor hold prior to any readout, and also on the Accessibility Proposition in Part Zero. This part will take up this argument directly rather than evade it.

Within this framework, the present part unfolds around four interconnected topics.

First, probability is not a new fact, but the result of truncating generative information. The core conclusion of this part, especially Chapter 18, is that probability is not a fundamental property of nature, but a structural truncation of the generative process by the readout layer. Gating, bandwidth, phase fidelity, and threshold are not engineering details, but institutional conditions determining which generative information can enter the recordable world. Once these conditions exist, phase information is inevitably and irreversibly lost, and probability appears as a consequence. Probability is therefore neither "ignorance" nor "ontic uncertainty," but a degraded invariant of generative information at the readout layer.

Second, probability must be adjustable; otherwise, the theory cannot hold. If probability indeed originates from readout structure, then it must respond systematically to changes in the parameters of that readout structure. On this basis, this part proposes a strong requirement: probability distributions must vary systematically with gating, bandwidth, and phase fidelity. This requirement is not an additional assumption, but transforms probability from an explanatory tool immune to failure into a physical object that can be pressured and negated by experiment. In this sense, this part is not merely "explaining probability," but setting failure conditions for probability—this is precisely the empirical collateral of the Accessibility Proposition in Part Zero.

Third, interaction is not force, but the rewriting of generative constraints. This part, especially Chapter 17, pushes this logic further: in the generative framework there is no "force acting on objects." What is called interaction is the mutual prehension and rearrangement of generative constraints among frozen structures. Its consequence is not that "an action has been exerted," but that the freezing budget is redistributed, the stability window changes, and the set of closeable paths is systematically rewritten. What the readout layer then sees are familiar phenomena such as scattering, deflection, delay, and changes in lifetime.

Fourth, why traditional forms remain effective. This part does not deny the language of traditional physics, but specifies its layer of applicability: force is the slope of readout response; potential is the effective representation of the distribution of freezing budget; field is the continuous approximation of the density of frozen structures; and action extremization is the statistical optimum of closeable-path density. These forms are successful not because they reveal the ontology of force, but because they calculate the statistical geometry of generative constraints.

Within the structure of the whole book, Part IV plays an interfacing role: it connects upstream to generative ontology, closure dynamics, and frozen facts, and downstream to entropy, thermodynamics, macroscopic irreversibility, and eventually gravitation and the continuous limit of spacetime. Only after probability and interaction have been explicitly moved down to the readout layer can macroscopic physics and the problem of gravitation avoid contaminating the generation layer in reverse. This part therefore unfolds in five chapters: Chapter 14 argues why the readout layer must be introduced, and directly addresses the two key concerns of "dissolution rather than relocation" and the "pre-observer universe"; Chapter 15 gives the readout origin of time, space, and causality, while identifying structural tensions not yet crossed; Chapter 16 rewrites particles, mass, and stable structures; Chapter 17 liberates interaction from "force"; and Chapter 18 provides the complete mechanism and falsifiable criteria for probability as a truncation phenomenon.

In short, Part IV does not introduce new existents, but completes a necessary accounting: it withdraws probability, interaction, force, and randomness from the ontological language of the world and repositions them as appearances left by generative facts at the readout layer. Only after this step is completed does physics acquire the logical qualification to enter the macroscopic and cosmic scales.

14. Why the Readout Layer Must Be Introduced

The preceding three parts established a complete generation layer: events, relations, closure, freezing, and the institutional principles they must obey. But what this generation layer produces are determinate, discrete, settled frozen structures, whereas the world we actually experience is continuous, statistical, and filled with the appearance of "interaction." There is a gap between the two that must be taken seriously. This chapter introduces the readout layer in order to locate this gap. It aims to establish that generation is not observation, and that the readout layer is a necessary rupture rather than a patch (14.1); that "seeing" and "measuring" are not physical verbs, and readout requires no observer (14.2); that the readout protocol consists of the triple truncation of bandwidth, gating, and threshold, from which the probabilistic appearance inevitably arises (14.3); that "state space" is readout space rather than generation space (14.4); and that changing the readout protocol changes the presentation of "reality," which is the core falsification path of this book (14.5). This chapter will also directly answer two of the sharpest objections: whether the readout layer merely renames and relocates the measurement problem, and whether the pre-observer universe thereby becomes inexplicable.

14.1. Generation $\neq$ Observation

"Generation is observation" is a fatal slide. Once "being generated" is equated with "being observed," the generation layer is immediately contaminated by the readout layer: existence becomes dependent on observation, while observation presupposes an observer, a moment of observation, and a background in which observation occurs.

Container grammar thereby reenters through the back door. Generation and observation must therefore be strictly separated in logic: generation is the production of facts, while observation is the lossy acquisition of facts. The two belong to different layers and perform different functions.

What would inevitably happen if the readout layer were not introduced? Generative theory would be forced to choose between two equally unacceptable options: either writing the statistical appearance of the world directly into the generation layer, in which case probability would be re-ontologized and the entire diagnosis of Chapter 3 would be overturned; or denying that the world has a statistical appearance, which directly conflicts with experience. The introduction of the readout layer is precisely what avoids this dilemma—it provides a functionally clear interface between the generation layer, which is determinate and discrete, and the experienced world, which is statistical and continuous. Therefore, the introduction of the readout layer is not a patch, but a necessary rupture: it is not temporarily added to repair some defect in generative theory, but explicitly acknowledges a layer difference that already exists between the generation layer and the experienced world. For the same reason, the language of "projection" must be abandoned: "projection" suggests that a higher-dimensional reality is projected onto a lower-dimensional display, with the two being different presentations of the same object; but the generation layer and the readout layer are not two presentations of the same object—readout is not a projection of generation, but a lossy sampling of generative results.

This leads to an objection that must be answered directly and cannot be dismissed merely by saying that the two layers have different functions: is the introduction of the readout layer merely renaming and relocating the measurement problem? That is, does this book merely rename the old question of "why measurement produces a definite result" from "collapse" in quantum mechanics into "readout" and move it to another architectural layer, without actually solving it? This objection is sharp because merely distinguishing the generation layer from the readout layer is not by itself enough to dissolve the measurement problem. A theory can distinguish two layers while still leaving "how readout produces a definite result" as an unresolved mystery; in that case, the readout layer would merely be a new name for collapse, and the problem would have been relocated rather than dissolved.

The answer of this book is: the measurement problem is dissolved, not relocated; but this answer has a clear dependency condition that the book openly acknowledges.

The criterial basis of dissolution lies in the settled predecessor being prior to readout. The sharpness of the measurement problem in standard quantum mechanics stems from a kind of constitutive inversion (Section 8.3): measurement seems to participate in constituting the very result it is supposed merely to reveal—collapse is both "reading out" and "causing," so "why and how measurement selects a definite result" becomes an insoluble puzzle. This book can dissolve rather than relocate the problem precisely because it cuts off constitutive inversion at the criterial layer: according to Section 8.3, closure is a genuine settlement criterion, and the settled predecessor it marks holds prior to any readout. Readout acts only upon this prior structure, which has already been adjudicated as complete by the generative process itself, and never participates in constituting it. Thus, in this book there is no such thing as "readout selecting the result." The result—the settled frozen structure—has already been adjudicated by the generation layer before any readout occurs. What readout does is only lossy sampling of this established fact under finite bandwidth, gating, and thresholds (Section 14.3). The question "why measurement produces a definite result" is thereby dissolved: the definite result is not produced by measurement, but already settled by the generation layer; measurement, or readout, only acquires it lossily. This is essentially different from relocation: relocation would make "how readout selects a result" replace "how collapse selects a result" as a new mystery, whereas dissolution cancels the step of "readout selecting a result" altogether.

However, this book must honestly mark the following point: the above dissolution depends on the Accessibility Proposition established in Section 0.3.4—the generation layer not only exists, but is at least partially accessible, meaning that it can be traced and its settlements can in principle be resolved. The reason is this: if the generation layer is constitutively inaccessible to direct description—if there exists a closed generation layer and readout is the only layer at which anything is given—then the statement that "the settled predecessor holds prior to readout" may be ontologically true, but it cannot be epistemically fulfilled as an explanation of measurement results. We would be unable to say what the "prior result" sampled by readout actually is, or why it is that result. Under such a closed reading, the measurement problem would indeed be relocated rather than dissolved. Therefore, this book claims dissolution, but the strength of this claim is exactly equal to the strength of the Accessibility Proposition: if accessibility holds, with its empirical collateral being the fourth item of "How to Falsify This Theory," namely the degradation law of readout modulation, then the measurement problem is dissolved; if accessibility is falsified, for example if statistical distributions do not degrade when phase fidelity is increased, then the book's claim of dissolution degenerates into relocation, and the anti-ontic-probability program fails together with it. The book makes this dependency entirely explicit rather than concealing it. This is the honest way to answer the objection: not by asserting that "we have completely solved the measurement problem," but by specifying precisely under what experimentally testable conditions the measurement problem is dissolved rather than relocated.

The functional boundary of the readout layer can now be stated clearly: readout is not explanation, but interface. The readout layer does not undertake the task of "explaining why existence is thus"; it undertakes interface definition, sampling limitations, information truncation, and gating rules. The generation layer answers "how the world becomes," while the readout layer answers only "how we can only acquire the world in this way." This is the legitimacy of the readout layer—it is an interface, not another layer of dynamics that must explain how results are produced. This section therefore establishes the basic position of Part IV: the generation layer and the observation layer cannot be logically identical—their functions do not collapse; without introducing the readout layer, generative theory would inevitably fall back into the old paradigm; readout is not projection, but output under the constraints of gating, bandwidth, and threshold; and the form in which the world appears belongs to the readout layer rather than the generation layer. This section has also directly answered the deepest objection: the readout layer dissolves, rather than relocates, the measurement problem—its basis is the settled predecessor prior to readout, its condition is the Accessibility Proposition, and both are explicitly marked.

14.2. Why "Seeing" Is Not a Physical Verb

Once the readout layer is introduced, its relation to the "observer" must be clarified immediately; otherwise, it can easily be misread as a form of observer-centered theory. This section aims to show that "seeing" is not a physical verb. It is a human special case within the readout layer, whereas readout itself requires no observer.

First ask: to which layer does "seeing" logically belong? "Seeing" implies three unaudited premises: there exists a subject that sees, there exists an object that is seen, and there exists a moment and background in which seeing occurs. All three premises belong to the readout layer, and specifically to a highly specialized form within it, namely human vision. To treat "seeing" as a verb describing the basic operation of the world is to elevate a highly specialized readout case into a fundamental operation of the universe. In generative logic, therefore, "seeing" cannot serve as a primitive operation: there is no subject, no object being seen, and no background in which seeing occurs at the generation layer; generation is only the adjudication of closure and freezing. "Measurement" is likewise not a physical verb: measurement presupposes a measuring device, a measurement setting, and a moment of measurement, all of which are specialized structures of the readout layer. At the generation layer, there is no operation called "measurement," but only the lossy mutual prehension among frozen structures. "Seeing" and "measurement" are therefore human languages, not the language of the universe. They describe how readout systems like us acquire the world, not how the world operates. Their only legitimate position is as "output" within the readout layer: a readout produces an output under a finite protocol, and "seeing" or "measuring" is the humanized description of such an output.

This leads to a seemingly fatal question: if the appearance of the world is assigned to the readout layer, does the pre-observer universe become inexplicable? Before any observer or measuring apparatus existed, the universe had already evolved for a long time—stars formed, elements were synthesized, and structures grew. If "appearance" belongs to the readout layer, and readout seems to require some kind of observer, would the pre-observer universe not become inexplicable? Does the book thereby slide into instrumentalism or observer-centered theory in which "without observation there is no world"?

This question is based on a misunderstanding that must be corrected: it assumes that "readout" is equivalent to "being seen by a macroscopic observer." In this book, the readout layer has no necessary relation to observers, consciousness, or measuring devices. The preceding discussion has already shown that "seeing" is only one human special case of readout. The general definition of readout is a lossy sampling protocol applied to frozen structures under finite bandwidth, finite time window, and finite resolution (Section 14.3). Such sampling can be realized by any physical process with finite resolving capacity and requires no observer, consciousness, or instrument at all—a beam of light passing through the matter distribution of the early universe, a particle scattering, or the finite response of one frozen structure to another frozen structure are all readouts, and all occur without any observer.

The more fundamental answer lies in the distinction between layers itself: the evolution of the early universe is a fact of the generation layer, not a product of the readout layer. According to Sections 8.3

and 12.2, the status of a settled, or complete, frozen structure is prior to and independent of any readout: closure is a genuine settlement criterion, and completeness holds only when the answer is yes to the question, "If the external world did not exist at all, could this generative process still stop self-consistently?" Therefore, the formation of stars and the synthesis of elements, as the generation and accumulation of frozen structures, occurred objectively and determinately at the generation layer, and their reality is not diminished by the absence of observers at that time. This directly resolves the unacceptable consequence rejected in Section 4.4, namely "no observation, no fact": facts are established by the settlement criterion of the generation layer, while observation, or readout, is only the later lossy acquisition of established facts. The position of this book is therefore the opposite of instrumentalism: it maintains a fully objective sequence of facts at the generation layer that depends on no observer. The readout layer does not create these facts; it only acquires them under finite protocols. Nor does "acquisition" itself require an observer; it requires only any physical process with finite resolution. The pre-observer universe therefore poses no difficulty: it is a fact of the generation layer, holding before any readout and before any observer. The introduction of the readout layer does not hand the world over to observers, but explains why we, as finite-resolution readout systems that appear later, can only acquire an already objectively existing generative history in forms such as "time," "space," and "probability." This section therefore establishes a conclusion that has institutional force for Part IV: "seeing" and "measurement" arc not physical verbs, but the language of the readout layer; without introducing the readout layer, they would illegitimately invade the generation layer; all observational descriptions must be rewritten as "readout output"; the world is not seen, but sampled and output.

14.3. The Readout Protocol: Bandwidth, Gating, and Threshold

Since readout is lossy sampling, the exact structure of this sampling must be specified. This section provides the three components of the readout protocol—bandwidth, gating, and threshold—and shows that it is precisely this triple truncation that makes determinate generative structures appear probabilistic at the readout layer.

A readout protocol is the set of limitations followed by any finite-resolution physical process when sampling frozen structures. It contains three indispensable elements. The first is bandwidth: any readout can respond only to rhythmic structures within a finite frequency band; components above or below that band are naturally filtered out. The direct consequence of bandwidth is the necessary loss of phase information—the complete phase relations at the generation layer cannot be fully preserved after band-pass filtering. The discarded phase details do not disappear, but appear as statistical dispersion. The second is gating: readout is effective only within a finite time window, while information outside the window is ignored. Gating makes readout necessarily discrete—it cannot continuously track phase evolution, but can only return whether a trigger occurs within a finite window. The third is threshold: readout responds only to structures that cross a certain threshold, while those below it are not recorded. Threshold makes "events" appear as jumps at the readout layer—generation that advances continuously in appearance is expressed, after threshold gating, as the discrete jump of "recorded or not recorded."

Once the three are combined, it becomes clear why the probabilistic appearance inevitably arises: bandwidth loses phase, gating discretizes time, and threshold discretizes response. Under this triple truncation, the determinate phase structure of the generation layer is compressed into a lossy and statistically dispersed output, and the distribution of this output is precisely what we call "probability." Probability is therefore not an assumption, but readout degradation: it is not an ontological component additionally introduced into the generation layer, but the statistical residue necessarily left when a determinate generative structure undergoes triple truncation. The conclusion of this section is: the readout protocol consists of the triple truncation of bandwidth, gating, and threshold; the probabilistic appearance is the necessary product of this triple truncation acting upon determinate generative structure, not the intrinsic randomness of the world. The full formalization of this mechanism—including the generation-layer operator counterparts of the three truncations and why $|\psi|^2$ appears as the invariant profile after truncation—will be given in Chapter 18.

14.4. The Illusion of Projection: Why "State Space" Is Not the World

The introduction of the readout layer requires us to reconsider a central object in quantum theory: state space, or Hilbert space. This section aims to show that state space is readout space rather than generation space, and that treating it as the world itself is an illusion of projection.

State space is readout space, not generation space. The structures of Hilbert space—superposition, basis, inner product, and projection—are all organizational forms of the readout layer. They describe how, under a given readout protocol, the sampling outputs of an already frozen structure can be organized, rather than how facts are produced at the generation layer. Treating state space as the world itself is to mistake readout space for generation space. This clarifies where "projection" actually occurs: it does not occur at the generation layer, where there are only closure and freezing and no operation called "projection." It occurs at the readout layer: the sampling of a frozen structure under a particular readout basis is mathematically represented as "projection." This also explains why "collapse" is not an event. In standard quantum mechanics, collapse is treated as a physical event occurring at some time and place. In this book, there is no such step as "collapse" at all—the result has already been settled by the generation layer before readout (Section 14.1). What is called "collapse" is only the mathematical appearance, in state-space representation, of a lossy sampling by readout of an established result, not a physical process that really occurs.

It must be acknowledged, however, that state space remains effective—just as continuity, force, and energy remain effective at the readout layer. State space is effective because it correctly captures the statistical structure of readout outputs. Its success lies not in revealing the ontology of the world, but in being an appropriate mathematical language for readout-layer sampling statistics. This book therefore does not discard state space, but reinterprets it from "state projection" to "readout mapping": a quantum state is not the state of the world, but an encoding of the sampling output of a frozen structure under a given readout protocol; the Born rule is not the probability law of the world, but the statistical weight of this readout mapping. The conclusion of this section is: state space is readout space, and treating it as the world is the illusion of projection. It is entirely effective at the readout layer, but has no ontological priority—the world is not in state space; the world is at the generation layer, and state space is only one statistical language by which we read it.

14.5. Readout Drift: Why Changing the Protocol Changes "Reality"

If the appearance of the world belongs to the readout layer, then changing the readout protocol should change that appearance. This section establishes this implication as an operational and falsifiable proposition—readout drift—and shows that it is the core falsification path of this book, while also clarifying why it is not subjectivism.

What is readout drift? It refers to the following: while the facts at the generation layer remain unchanged, systematically changing parameters of the readout protocol—bandwidth, gating, phase fidelity, and threshold—will produce systematic and calculable changes in the statistical output. This is a direct consequence of the lossy-sampling nature of the readout layer: since the output is the result of sampling, changing the sampling protocol necessarily changes the output. It must immediately be clarified that readout drift is not subjectivism. It does not claim that "reality changes with the observer's will," because the readout protocol is not a subjective choice, but a physical sampling limitation. Changing the protocol is an objective physical operation, such as changing the bandwidth of a detector, not a psychological act. Moreover, the fact at the generation layer remains unchanged throughout the drift—what changes is only our mode of acquiring an unchanged generative fact, not that fact itself. Readout drift is therefore the opposite of the claim that "reality is subjectively constructed": it presupposes an objective and invariant generation layer, and precisely this invariant underlying layer makes the proposition that "different sampling of it gives different outputs" meaningful and calculable.

Drift has three basic directions: increasing phase fidelity should shrink the statistical distribution or reveal previously concealed phase structure; changing the gating window should systematically change counting statistics; and cutting or extending bandwidth should produce calculable drift in the distribution. These three directions are important because they constitute the core falsification path of this book. The book does not ask anyone to "believe in generative ontology." It proposes only an unavoidable requirement: readout statistics must change systematically with protocol paramcters. If experiments show that, after phase fidelity is significantly improved, after gating and bandwidth are explicitly changed, and under strict control of environmental and systematic errors, the statistical distribution remains strictly unchanged, then the conclusion is direct and unambiguous: probability is not the result of readout degradation, and the book fails on this point. This is a proposition that can genuinely be overturned by experiment. Its operational thresholds—including measurable proxies such as coherence time, visibility, and the CHSH parameter, together with order-of-magnitude thresholds—are fully stated in Chapter 18 and Appendix H, and belong to the same line of argument as the fourth item of "How to Falsify This Theory" in Part Zero and the Accessibility Proposition. This is fundamentally different from traditional interpretations: the Copenhagen interpretation treats probability as a non-adjustable postulate, while the many-worlds interpretation treats it as branch weight; neither predicts that changing the readout protocol will systematically change statistics. This book treats probability as adjustable output and thereby places it under direct experimental pressure. The conclusion of this section is:

readout drift is the operational form of the implication that changing the protocol changes the presentation of "reality." It is not subjectivism—the facts at the generation layer remain unchanged—but the most direct falsification path of this book. It transforms probability from an untouchable ontological stipulation into a physical object that can be brought down by experiment.

14.6. Summary of This Chapter

Chapter 14 has completed a structurally pivotal task in the book: after the institutional closure of the generation layer, it introduces a readout layer that has an entirely different function and cannot functionally collapse into the generation layer.

This chapter has established that generation is not observation; the readout layer is a necessary rupture rather than a patch between the generation layer and the experienced world; "seeing" and "measurement" are not physical verbs, but human special cases at the readout layer; the readout protocol consists of the triple truncation of bandwidth, gating, and threshold, from which the probabilistic appearance necessarily arises; state space is readout space rather than generation space; and changing the readout protocol systematically changes the statistical appearance of the world. More importantly, the chapter has directly answered two of the sharpest objections, and these two answers are decisive for whether Part IV can stand as a "readout-layer program." First, is the readout layer merely renaming and relocating the measurement problem? The answer of this chapter is: the measurement problem is dissolved rather than relocated—the criterial basis is closure as a genuine settlement criterion, which makes the settled predecessor hold prior to any readout and thereby cancels the constitutive inversion by which "readout selects the result." But this dissolution explicitly and testably depends on the Accessibility Proposition. The book marks this dependency honestly rather than concealing it. Second, does the pre-observer universe become inexplicable? The answer of this chapter is: readout is not the seeing of an observer, but the lossy sampling of any finite-resolution physical process. The early universe, as a fact of the generation layer, objectively holds prior to any readout and prior to any observer. The book is therefore precisely the opposite of instrumentalism: it maintains a fully objective sequence of generation-layer facts independent of observers.

The legitimacy, function, and boundary of the readout layer have thereby been established: it is an interface rather than dynamics, acquisition rather than creation, and lossy sampling rather than projection. The next chapter will, on the basis of this readout layer, provide the readout origin of time, space, and causality, while honestly marking the structural tensions not yet crossed.

15. The Readout Origin of Time, Space, and Causality

Chapter 14 established the function of the readout layer: it is lossy sampling of frozen generative structures. This chapter uses the readout layer to provide the generative origin of the three most basic "background" concepts—time, space, and causality. It aims to show that "earlier and later" is not a temporal coordinate, but the structural thickness of generative dependence (15.1); that "where" is not spatial position, but relational density (15.2); and that "earlier causing later" is not a dynamical postulate, but freezing reachability (15.3). But the chapter does not stop there: it also honestly marks four major structural tensions that have not yet been crossed on the path from this generative framework to full modern physics, and explicitly lists them as constructive tasks with criteria for success or failure (15.4). This final section is a concentrated expression of the book's discipline of honesty.

15.1. Temporal Depth: Why "Earlier and Later" Is Not a Coordinate but Structural Thickness

One cannot begin from a "time parameter." This is the position already established in Part II (Chapter 7). The task of this chapter is to implement that position at the readout layer through the operational concept of "temporal depth." The reason is that the time parameter t presupposes continuous, uniform, and reversible passage, and these are precisely the container structures that have been suspended. If one begins from t, time has been treated as a premise rather than a result.

Temporal depth is defined as the accumulated number of layers of frozen predecessors on which a frozen structure depends in generative history. It is not the reading of an external clock, but the thickness of generative dependence: the more layers of frozen predecessors a structure depends on, the greater its temporal depth. Accordingly, "earlier and later" comes from dependence rather than coordinates: to say that A is "before" B, in the exact sense used in this book, means that A, as a frozen predecessor, is depended upon by the generation of B, not that the time coordinate of A is smaller than that of B. Earlier and later constitute a partial order of generative dependence, not magnitudes on a coordinate axis. Temporal depth naturally has directionality, because dependence is one-way—the frozen is depended upon by subsequent generation, not the reverse. This directionality is precisely the expression, at the readout layer, of freezing monotonicity (Sections 7.4 and 11.4), and no additional arrow of time needs to be introduced.

Temporal depth must be distinguished from "duration": duration is a metric quantity at the readout layer, measured by some periodic process as a clock, whereas temporal depth is the dependency thickness of the generation layer. The former presupposes a clock; the latter presupposes no clock. The relation between temporal depth and observable time is therefore this: observable time is the metric appearance of temporal depth at the readout layer, obtained by means of some periodic frozen structure selected as a clock. Different clocks, meaning different periodic frozen structures, give different observable times, but all are readouts of the same thickness of generative dependence. The conclusion of this section is: time is not a coordinate axis, but the structural thickness of generative dependence; "earlier and later" is the partial order of freezing dependence; the directionality of time comes from the one-way character of freezing; and observable time is the readout of this thickness by means of clocks.

15.2. Spatial Distribution: Why "Where" Is Not Position but Relational Density

In parallel with time, "position coordinates" also cannot serve as a starting point: position presupposes a spatial container in which an event can "be located," and this is precisely what Chapter 1 suspended. This chapter implements space at the readout layer as "relational density."

Spatial distribution is defined as the density organization of relational weights among frozen structures. The "spatial" structure of a region is not its coordinate distribution on a presupposed manifold, but the density and orientation of relational weights among frozen structures in that region. Accordingly, proximity arises from coupling rather than distance: to say that two structures are "near" each other means, in the exact sense used in this book, that there is strong relational weight between them, making them more readily able to enter generation together, not that their coordinate distance is small. Distance is the readout of proximity, not the source of proximity. This is consistent with Section 6.4, where "the network is the source of space." Dimension is not a postulate, but statistical stability: the dimensionality of space is not a presupposed 3 or 4, but a stable statistical dimension appearing after a large number of directional relations have frozen. This is consistent with Section 5.3, where "direction precedes dimension." Under different frozen organizations, effective dimensionality can in principle be different. Three-dimensional space is the stable readout presented by the freezing of directional relations in our universe, not an a priori geometrical fact.

Precisely because space is the readout of relational density, it is necessarily "coarse." At a sufficiently fine scale, the appearance of continuous space gives way to a discrete relational network; what is called the "smoothness of space" is only a statistical appearance under finite readout resolution. This is consistent with Section 6.5, where continuity is the statistical appearance of freezing. Finally, the symmetry and asymmetry between space and time must be noted: both are readouts of generative structure, which is their symmetry; but temporal depth arises from the one-way partial order of freezing dependence and is therefore intrinsically directional, whereas spatial distribution arises from the undirected density of relational weights and therefore has no intrinsic direction. This asymmetry is the generation-layer source of why "time has an arrow but space does not." It requires no additional physical assumption, but follows directly from the structural difference that "dependence is one-way, while proximity is undirected." The conclusion of this section is: space is not a positional container, but the readout of relational density; proximity comes from coupling; dimensionality is statistical stability; and the continuity and smoothness of space are appearances of finite-resolution readout.

15.3. Causality: Why "Earlier Causing Later" Is Not a Dynamical Postulate

After time and space, the third background concept that must be rewritten is causality. The "law of causality" cannot serve as a fundamental postulate, because it presupposes the temporal order in which "cause precedes effect" and the spatial structure of "influence propagation"—but in this book time and space are both results at the readout layer (15.1, 15.2). To treat causality as a postulate is to smuggle readout-layer structures back into the generation layer as premises.

In this book, the only definable "earlier and later" is the order of freezing: the exact meaning of A "causing" B is that A, as a frozen predecessor, is depended upon by the generation of B. Causality is generative dependence, no more and no less. Accordingly, a "causal cone" is not a geometrical structure, but freezing reachability. The traditional light cone is a geometrical object in spacetime, whereas the "causal cone" in this book is the reachability structure of "which frozen structures can serve as predecessors of a given generation." It presupposes neither light speed nor spacetime, but is directly given by reachability relations at the generation layer. Its boundedness is precisely the discrete light-cone bound mentioned at the end of 15.2. This also explains why there is no problem of "retrocausality": because causality is freezing dependence, and freezing is monotonic and irreversible (Section 11.4), it is impossible by definition for a result to depend on a predecessor that has not yet frozen. Thus, "reverse causation" and "causal loops" are not anomalies that need to be prohibited in this book; they have never been legitimately introduced. They require dependence on something that has not yet become an acquired predecessor, and this has no definition in generative logic.

The relation between causality and probability is thereby clarified as well: probability does not define causality, as in some statistical theories of causation; rather, causality, or freezing dependence, precedes probability. Probability is the statistical appearance, at the readout layer, of a large number of frozen dependency structures (Chapter 18), while freezing dependence itself is determinate at the generation layer. Finally, operational causality must be clarified: in experiments, the causality we "measure" is the lossy sampling output, at the readout layer, of freezing dependence at the generation layer. What we observe as "A causes B" is the readout-protocol sampling of the generation-layer fact that "B depends on the frozen A." The conclusion of this section is: causality is not a dynamical postulate, but freezing reachability; the causal cone is a generative reachability structure rather than a geometrical light cone; retrocausality is an undefined pseudo-problem; and causality precedes probability rather than being defined by probability.

15.4. Work Not Yet Completed: Four Structural Tensions as Constructive Tasks

The preceding chapters of this book have reconstructed time, space, and causality as readout origins of generation–closure–freezing without presupposing any background. But it must be honestly acknowledged that several major structural tensions remain unresolved on the path from this generative framework to full modern physics. This section marks four of them and explicitly identifies them as constructive tasks yet to be completed, rather than questions that the book claims already to have answered. The position of this book is that an honest framework should clearly mark the places it has not yet reached, rather than covering gaps with more narrative. This is precisely the posture required by "How to Falsify This Theory" and Section 0.3.3.

First: Lorentz invariance. The formal compatibility between a discrete event-generation model and special relativity has not yet been rigorously handled in this book. The core difficulty is how discrete generative steps are to be reconciled with continuous Lorentz transformations, and how the discrete light-cone bound is coarse-grained at the readout layer into a continuous relativistic light cone and recovers Lorentz symmetry. This book does not claim to have derived Lorentz invariance. Its starting point is the discrete light-cone bound established in 15.2 and 15.3, denoted $c_{\text{gen}} = 1$, meaning that at the generation layer there exists a bounded propagation structure defined purely relationally and without presupposing spacetime—generative dependence at each logical step can advance only within a finite relational neighborhood. This purely generation-layer bounded propagation structure, which presupposes no spacetime, is the natural starting point for discussing Lorentz compatibility. One feasible research path is to show how the relativistic light cone appears as an effective readout-layer approximation of the discrete light-cone bound. But the derivation from "discrete light-cone bound" to "complete Lorentz invariance" is a clearly unfinished task.

Second: the emergence of gauge symmetries U(1)/SU(2)/SU(3). This book openly treats the question of how the gauge structure of the Standard Model could arise from closure dynamics as an open problem. The tendency of directional extension at the generation layer (Section 5.3) and the directedness of weights (Section 6.6) provide a kind of intrinsic orientational freedom, which may in principle be the generation-layer predecessor of gauge freedom. But this book provides no derivation, and does not pretend to provide one, from the triadic intrinsic structure to the concrete gauge groups U(1), SU(2), and SU(3). This is a step that the framework must cross in order to move toward the Standard Model of particle physics, and it has not yet been crossed.

Third: renormalization and ultraviolet divergence. Discreteness can in principle regularize ultraviolet divergence—since the generation layer is discrete and contains no continuous infinite divisibility (Chapter 2), the continuous field-theoretic assumptions that lead to ultraviolet divergences do not hold at the generation layer in the first place. But this point must be proven, not assumed. More specifically, it must be explained how the ultraviolet behavior of effective field theory at the readout layer (see the hook at the end of Chapter 22 and the EQT bridge in Appendix I) is regularized by the discreteness of the generation layer, and whether this regularization naturally yields finite physical quantities. At present, the book can only point out that discreteness provides a principled possibility of regularization, and lists "how discreteness concretely regularizes readout-layer field theory" as a task yet to be completed.

Fourth: whether the measurement problem is genuinely dissolved. This item differs in nature from the preceding three—it is not "not yet derived," but "dependent on an explicitly marked proposition." Section 14.1 has argued that this book claims the measurement problem is dissolved rather than relocated, but this dissolution depends on the Accessibility Proposition in Section 0.3.4. If accessibility is falsified, dissolution degenerates into relocation. Therefore, "whether the measurement problem is genuinely dissolved" is not a closed conclusion in this book, but a proposition whose condition of validity has been explicitly marked and whose condition itself is experimentally falsifiable. The empirical collateral is the fourth item in "How to Falsify This Theory," the degradation law of readout modulation. This book lists it here in order to present all major unresolved tensions of the framework in one place and to preserve honest completeness.

A methodological clarification is needed regarding these four markings. Listing these four items is not an admission that the book has failed, but a fulfillment of its core discipline: a theory that cannot fail does not deserve acceptance (Section 0.3.3). Each of these four items provides a clear criterion by which progress can be tested: whether Lorentz invariance can be recovered from the discrete light-cone bound; whether gauge groups can emerge from directional freedom; whether ultraviolet divergences can be regularized by discreteness; and whether the accessibility degradation law can be experimentally observed. They are therefore not vague "future work," but constructive tasks with clear standards of success or failure. This book would rather clearly mark these four thresholds that have not yet been crossed than conceal them with narrative as "solved in principle." The latter is precisely the "always just short of completion" grammar of deferral diagnosed and rejected in Part I.

15.5. Summary of This Chapter

Chapter 15 has completed the core construction of Part IV: time, space, and causality are not presupposed background structures, but the results of generation–closure–freezing at the readout layer.

This chapter has established that time is not a coordinate, but the structural thickness of generative dependence; "earlier and later" is the partial order of freezing dependence, and the directionality of time comes from the one-way character of freezing. Space is not a positional container, but the readout of relational density; proximity comes from coupling, dimensionality is statistical stability, and continuity and smoothness are appearances of finite-resolution readout. Causality is not a dynamical postulate, but freezing reachability; the causal cone is a generative reachability structure rather than a geometrical light cone, and retrocausality is an undefined pseudo-problem. Together, the three show that at the generation layer there are only freezing dependence and relational density; time, space, and causality are organizational appearances of these generative structures at the readout layer, not a priori stages within which generation unfolds.

But perhaps the most important work of this chapter lies in its final honest marking: from this generative framework to full modern physics, four major structural tensions remain uncrossed—the recovery of Lorentz invariance, the emergence of gauge groups, the discrete regularization of ultraviolet divergences, and the Accessibility Proposition on which dissolution of the measurement problem depends. This chapter explicitly lists them as constructive tasks with criteria for success or failure, rather than concealing them as "solved in principle." This marking is not a weakness, but a fulfillment of the book's core discipline: only a framework that clearly marks where it has not yet reached is capable of failure and therefore worthy of serious consideration as a physical framework. The next chapter will, on the basis of the readout origin of time, space, and causality, turn to how particles, mass, and stable structures appear as readouts of frozen structures.

16. Particles, Mass, and Stable Structures

Chapter 15 provided the readout origins of time, space, and causality. This chapter continues that work by addressing three concepts regarded as fundamental in traditional physics, yet which can likewise be rewritten as readouts of freezing: particles, mass, and stable structures. It aims to show that a particle is not an object, but a readout of a frozen structure (16.1); that mass is not an attribute of an entity, but a readout of the freezing budget (16.2); and that stable structures are not the norm, but the result of extreme filtering and spectral sparsity (16.3). In particular, the section on mass will establish a discipline that runs throughout the book and its appendices—absolute mass is a calibration, whereas the mass difference Δm bears the genuine responsibility of falsifiable prediction. On this basis, it provides the mechanism chain for the proton–neutron mass difference, which is one of the core points at which this book connects with high-precision experiment.

16.1. A Particle Is Not an Object, but a Readout of a Frozen Structure

The idea of the "particle as a fundamental object" cannot hold in this book, for reasons already given repeatedly (Chapter 4): an object presupposes persistent identity and prior existence, whereas at the generation layer there are no prior objects, only generative adjudications. This chapter implements that position positively at the readout layer by rewriting the concept of the "particle."

At the generation layer, the only fact capable of persistence is a frozen structure: a set of events successfully closes, is repeatedly confirmed, and freezes, thereafter participating in subsequent generation as an acquired unit (Chapters 12 and 13). This is the only thing at the generation layer that possesses persistence. How does the readout layer "see" a particle? When a stable frozen structure is sampled by a finite-resolution readout protocol, it appears as a "thing" with determinate properties—mass, charge, spin, and other readout quantities—that can be repeatedly identified. This read-out stable unit is what we call a "particle." A particle is therefore not a fundamental entity at the generation layer, but the stable appearance of a frozen structure at the readout layer. Accordingly, "particle number" is a readout variable rather than an ontological conserved quantity: under different readout protocols, the same generative structure may be sampled as different "particle numbers"—consistent with the fact in quantum field theory that particle number depends on the choice of vacuum and on the observer's state of motion. The "conservation" of particle number is a statistical regularity under particular readout conditions, not an ontological conservation law at the generation layer. The true source of particle stability is the freezing depth and defreezing threshold of its corresponding frozen structure (Section 9.2): a "stable particle" corresponds to a repeatedly confirmed frozen structure with a high defreezing threshold, whereas an "unstable particle," or decay, corresponds to a frozen structure with a lower defreezing threshold that will eventually be reorganized in higher-level restructuring. The conclusion of this section is: a particle is a readout of a frozen structure, particle number is a readout variable, and particle stability is a readout of freezing depth—there is no ontological category of "fundamental particles," only stable frozen structures that are read out.

16.2. Mass Is Not an Attribute of an Entity, but a Readout of the Freezing Budget

In traditional physics, "mass" is almost inevitably understood as an intrinsic attribute of an object: an object "has" mass, a particle "carries" mass, and a field "gives" mass. Even in modern quantum field theory, where mass is reformulated as a parameter in the Lagrangian or as a result of Higgs coupling, this attributive intuition has not truly been removed. This book adopts a fundamentally different position: mass is not an attribute at the generation layer, but the aggregate readout of the "budget" required by a frozen structure.

Before developing the generative definition of mass, a methodological discipline running throughout this chapter and Appendices C and F must first be established; otherwise, readers may easily misread the specific mass values appearing below and in the appendices as numerical fitting. This book explicitly distinguishes two tasks of entirely different kinds.

The first is absolute-mass calibration. To map the freezing budget at the generation layer—a dimensionless scalar, or one expressed in units of frequency—onto a specific mass value measured in MeV, an external readout mapping $m = \kappa, \mathfrak{m} + m_0$ must be introduced, in which the unit-mapping constant κ and zero point m_0 must be calibrated by external standards and cannot be supplied by generative dynamics itself. Therefore, any absolute mass value—for example, $m_p \approx 938.3$ MeV reported at the presentation scale in Appendix F—is not a prediction, but a calibration. It depends on the externally selected κ, m_0, and presentation energy scale. This book openly acknowledges this point: specific absolute-mass values do not bear the falsifiable responsibility of the book. In particular, initializing the central frequency as $f_c = 930$ in Appendix F is merely an energy-scale anchor for presentation, used to report frequencies directly on the MeV scale. It is not a predicted quantity and does not constitute a dynamical input. The resulting value of 938.3 should therefore be understood only as "a consistency check at this presentation scale," not as an independent prediction. To criticize $930 \to 938.3$ as fitting an input close to the output is precisely to misplace the level of the book's claim—the book has never treated absolute mass as a prediction.

The second is the differential output Δm. What genuinely bears falsifiable responsibility in this book is the mass difference, typically the proton–neutron mass difference. The key point is that, in a difference, the external zero point m_0 is structurally eliminated, and sensitivity to common calibration terms is also substantially reduced (see Appendix C). Therefore, Δm does not depend on the specific choice of absolute calibration, but is directly determined by the mechanism chain at the generation layer. Δm is the genuine prediction made by this book concerning mass, and the object subject to high-precision experimental testing and falsification. Accordingly, in the discussion of mass below, the emphasis is placed on the mechanism chain of Δm, rather than on absolute mass values. Absolute mass serves only as an example of how a freezing budget is read out after calibration, and its numerical value bears no predictive responsibility.

Under this discipline, the generative definition of mass can now be given. There is no "mass variable" at the generation layer: the only permitted basic quantities at the generation layer are the triadic intrinsic structure of events, relational weights among events, and the criteria of evolution–closure–freezing. At this layer there is no "object," no "energy carrier," and no logical position for "mass conservation." To forcibly introduce a mass variable at the generation layer would simultaneously assume that stable objects already exist, that comparable "inertia" already exists, and that a unified readout scale already exists. All three assumptions violate the principles of no a priori objects, no a priori scale, and no a priori spacetime. The generative definition of mass therefore begins from the "freezing budget": for a frozen structure to hold, it must continuously satisfy a set of conditions—the frequency configuration required for phase closure, the coupling strength required for relational weights to remain stable, and the generative cost of resisting disintegration and rearrangement.

Together, these conditions point to one fact: freezing a structure has a cost. This cost is neither "energy consumption" nor "the amount of work performed," but the total quantity of constraints that must be continuously satisfied in generative dynamics. This book calls it the freezing budget. When a readout protocol is introduced, the freezing budget is mapped onto a scalar output; under a specific and stable readout mapping, this output is labeled "mass." Thus, greater mass does not mean containing more "matter" or possessing stronger entityhood, but requiring a larger generative budget to maintain the frozen structure. Mass is structural: it is a compressed readout of the global generative cost, rather than any local and portable attribute. It must be emphasized that the readout mapping that "maps the freezing budget onto a scalar output" is precisely the mapping $m = \kappa\mathfrak{m} + m_0$ on which "absolute-mass calibration" in the preceding discipline depends. Therefore, the absolute mass readout of a single structure depends on external calibration and bears no predictive responsibility. Only when the difference between the budgets of two frozen structures is compared does the external zero point cancel, yielding a genuinely mechanism-determined and predictable output.

This leads to the core of this section: the mechanism chain of the mass difference Δm. According to the preceding discipline, the genuine prediction made and exposed to falsification by this book concerning mass is the mass difference rather than absolute mass. Taking the proton–neutron difference as an example, its generative mechanism is not a single peripheral compensation, but the sum of several same-sign frequency differences. Under the narrow-band approximation, the proton–neutron mass difference can be written as

$$\Delta M = \frac{h}{c^2}\left[,\Delta\bar{f}_g + (f_d - f_u) + O(\sigma_f^2),\right],$$

where the sources and status of the three terms differ and must be explained separately—in particular, the dominant term without free parameters must be distinguished from the estimated small correction term.

The dominant term is the flavor difference $(f_d - f_u)$. It contains no free parameters and independently bears falsifiable responsibility.

In this model, quark flavor is distinguished by event phase: up-type events correspond to $\theta = 0$, while down-type events correspond to $\theta = \pi$, consistent with phase as the internal state of rhythm in Section 5.2. The peripheral phases of the proton configuration (uud) are $(0,0,\pi)$, whereas those of the neutron configuration (udd) are $(0,\pi,\pi)$—the neutron contains one more peripheral event with $\theta = \pi$ than the proton. The flavor-difference term $(f_d - f_u)$ is precisely the frequency difference between down-type and up-type peripheral events. It is a purely peripheral frequency difference, contains no central anchor, and requires no fitted parameter. Substituting the original numerical value $f_d - f_u = 3.13 \times 10^{20}$, Hz, this term alone yields approximately 1.29 MeV—already essentially consistent with the experimental value of 1.293 MeV. In other words, the main body of the proton–neutron mass difference is independently carried by a parameter-free peripheral flavor-difference term.

The correction term is the central frequency shift $\Delta\bar{f}_g$. It is an estimated small correction whose sign is determined by the mechanism, but whose numerical value is estimated. The neutron's two peripheral events with $\theta = \pi$ introduce greater phase tension—opposite phase increases the cost of maintaining synchronization, as discussed in Sections 8.2 and 19.3. Through nonlinear feedback, this tension slightly raises the mean freezing frequency of the central event, so $\Delta\bar{f}_g > 0$. Its sign is determined by the mechanism: the neutron has greater phase tension and therefore a higher central freezing frequency, giving a positive contribution to ΔM, with the same sign as the flavor-difference term. This is why $\Delta M > 0$, meaning that the neutron is heavier, follows naturally rather than through fine-tuning. But this book must honestly mark that the specific numerical value of $\Delta\bar{f}_g$ is estimated rather than derived from first principles. The original value $\Delta\bar{f}_g \approx 0.5 \times 10^{20}$, Hz, approximately 0.21 MeV, is an estimated small correction, and its falsifiable responsibility is weaker than that of the flavor-difference term. The book does not conceal this point: the hard core of the mass difference is the parameter-free flavor-difference term, while the central term is an openly acknowledged estimated correction. Adding the two terms gives $\Delta M \approx 1.29 + 0.21 \approx 1.5$ MeV, reported under the original rounding convention as $\Delta M \approx 1.3$

MeV, with a positive sign, meaning that the neutron is heavier. The scale is consistent with experiment, and the result is obtained without fine-tuning, arising naturally from "flavor-difference dominance + a small central correction induced by phase tension."

The relation between ΔM and the central anchor f_c, approximately 930, must be made clear here, because it is crucial to answering the objection of "fitting numbers to numbers": ΔM is independent of the anchor. The energy scale of the central event, $f_c \approx 930$, determines the dominant term of the absolute baryon mass—approximately 98% of baryon mass comes from the central event, consistent with the QCD view that baryon mass is dominated by gluon-field energy. Thus, $M_{\text{baryon}} \approx hf_c/c^2 + (\text{small peripheral terms})$. But no term in ΔM contains the additive value of f_c: the flavor-difference term is a peripheral frequency difference, while the central correction term is the central frequency difference $\Delta\bar{f}_g$, not its absolute value. Therefore, f_c enters absolute mass but not ΔM. The direct consequence is that ΔM, in absolute units, is independent of the choice of the central anchor f_c. This will be demonstrated numerically in Chapter 20: when f_c ranges from 600 to 1500 MeV, ΔM remains approximately 1.3 MeV. The relation $930 \to 938$ concerns the scale of absolute mass, whereas approximately 1.3 is determined by an independent flavor-difference term. Their origins are different. To challenge ΔM by treating the former as "choosing an input to match the output" is precisely to misplace the mechanism—ΔM does not read the value 930 at all. Finally, falsifiability must be stated explicitly: the dominant contribution to ΔM from the flavor-difference term bears the responsibility of falsification. It contains no free parameters, is determined by the peripheral phase configurations (uud) versus (udd) and the flavor-frequency difference, and interfaces with high-precision measurement of the proton–neutron mass difference (see falsification path F1 in Appendix C). If this dominant term remains incompatible with experiment within the same calibration framework and cannot be repaired without violating conservation and freezing monotonicity (Chapter 11), then the mechanism is falsified. The central correction term, as an estimated quantity, is entered separately into the error budget (Appendix C) and is not used to "rescue" the dominant term.

This view of mass also naturally explains its other properties. The approximate additivity of mass is not an axiom, but a readout approximation under weak coupling among frozen structures: when two frozen structures combine, their internal closure modes remain mutually independent, and no significant structural rearrangement occurs, the freezing budgets add approximately linearly in the weak-coupling limit, appearing at the readout layer as additive mass. Under strong coupling, shared closure, or structural rearrangement, mass defects, nonlinear mass changes, and apparent "binding-energy" effects naturally arise without introducing negative energy or remedial mechanisms. Inertia is also repositioned: it is not "resistance to motion," but the resistance of a frozen structure to generative rearrangement. When the external generative environment changes, the frozen structure tends to preserve its existing closure mode; any rearrangement requires the freezing budget to be paid again, and therefore appears as "difficulty of change." At the readout layer, this resistance is interpreted as "inertial mass." Mass can also disappear, transform, or be redistributed: because mass is not an ontological attribute but a readout result, when a frozen structure disintegrates the mass readout disappears; when a frozen structure reorganizes, mass is redistributed; and when multiple freezing modes merge, mass changes nonlinearly. The appearances of particle annihilation, mass–energy transformation, and mass fluctuations near the "vacuum" thereby receive a unified interpretation: they are not the creation and destruction of entities, but the reconfiguration of freezing budgets within generative space. The conclusion of this section is: no mass variable exists at the generation layer; frozen structures require generative budgets; mass is a readout of the freezing budget; mass additivity is approximate rather than axiomatic; and inertia arises from resistance to generative rearrangement. This section also establishes the discipline concerning mass that runs throughout the book and its appendices: absolute mass is a calibration, dependent on external κ and m_0, and bears no predictive responsibility; $f_c = 930$ is only a presentation anchor and 938.3 only a consistency check at the presentation scale. The mass difference Δm, by contrast, is a prediction: the zero point cancels, the result is determined by the mechanism chain, and it can be falsified by high-precision experiment.

16.3. Stable Structures, Spectral Sparsity, and the Selective Appearance of the World

The final section addresses a frequently overlooked yet fundamental fact: why is the world composed of only a few kinds of stable structures rather than filled with arbitrary structures? The answer of this book is: stability is not the norm, but the result of extreme filtering; the spectrum of the world is sparse precisely because the closure conditions are exceptionally stringent.

Stability is not the norm, but the result of extreme filtering. Sections 8.5 and 9.5 established that failure is the default result of generation. It follows that structures capable of stable freezing are extremely rare—the vast majority of generative attempts cannot close, or close but cannot be repeatedly confirmed and therefore disintegrate. Stable structures are thus not things that "normally appear," but the extreme minority surviving among vast numbers of failures. This filtering is directly related to closure order and spectral sparsity: phase configurations capable of stable closure are discrete and highly constrained, because phase closure is a finite-step integer condition (Section 12.3). Therefore, freezeable structures are sparsely distributed across the "configuration spectrum," rather than continuously filling it. This explains why the particle spectrum is discrete and why only particular mass hierarchies appear, rather than arbitrary masses existing continuously.

It must be emphasized that stable structures are not "selected," but "permitted," consistently with the anti-teleological position of Section 9.1. There is no mechanism or purpose that chooses stable structures; there is only the threshold of the closure condition—those that pass are permitted to freeze, while those that do not pass fail. Stable structures are permitted survivors rather than selected winners. The stability of the world therefore comes from "repeated freezing," not "eternal existence": a stable structure persists not because it becomes eternally invariant once generated, but because its closure mode is repeatedly reconfirmed in subsequent generation (Section 12.4). Stability is dynamic repeated success rather than static eternity. Spectral sparsity is also related to the "intelligibility" of the world: precisely because freezeable structures are sparse, discrete, and classifiable, the world presents a finite spectrum that can be recognized, classified, and named by finite readout systems. If every structure could remain stable, the world would be an undifferentiated saturated whole, incapable of classification or understanding, as discussed in Section 13.3. Particle spectra, mass hierarchies, and sparsity thereby receive a unified explanation: they are different aspects of the same fact—the discreteness and stringency of closure conditions cause freezeable structures to be sparsely distributed across the configuration spectrum, appearing at the readout layer as discrete particle types and discrete mass hierarchies. The conclusion of this section is: stable structures are extreme survivors of filtering rather than the norm; spectral sparsity arises from the discreteness and stringency of closure conditions; and the selective appearance of the world—only a few structures exist—and its intelligibility—the world can be classified—are two sides of the same sparsity.

16.4. Summary of This Chapter

Chapter 16 has uniformly rewritten three concepts regarded as fundamental in traditional physics—"particle," "mass," and "stable structure"—as results of generation–closure–freezing at the readout layer.

This chapter has established that a particle is not a fundamental object, but the appearance at the readout layer of a stable frozen structure; particle number is a readout variable, and particle stability is a readout of freezing depth. Mass is not an attribute of an entity, but a readout of the freezing budget. In the section on mass, the chapter established a key discipline running throughout the book and its appendices: absolute mass is a calibration, dependent on the external mapping $m = \kappa \mathfrak{m} + m_0$, and its value bears no predictive responsibility; $f_c = 930$ is only a presentation anchor. The mass difference Δm, by contrast, is the genuine prediction: the zero point cancels, the result is determined by the mechanism chain, and it can be falsified. On this basis, the chapter provided the mechanism chain for the proton–neutron mass difference: its main contribution is the parameter-free flavor-difference term $(f_d - f_u)$, which alone yields approximately 1.29 MeV, essentially consistent with the experimental value of 1.293 MeV, and independently bears falsifiable responsibility. The central frequency shift $\Delta \bar{f}_g$ is a small correction whose sign is determined by the mechanism but whose numerical value is estimated, approximately 0.21 MeV; its uncertainty is entered separately into the error budget and is not used to rescue the dominant term. The two same-sign terms together yield $\Delta M \approx 1.3$ MeV, with the neutron heavier. Moreover, ΔM does not read the central anchor f_c and is therefore anchor-independent—this is the key response to the objection of "fitting numbers to numbers." Finally, the chapter provided a unified explanation of stable structures, spectral sparsity, and the selective appearance of the world as consequences of the discreteness and stringency of closure conditions: stability is the survival of extreme filtering rather than the norm, and the finite spectrum and intelligibility of the world are two sides of the same sparsity.

Particles, mass, and stable structures have thereby all been located at the readout layer, as readouts of frozen structures rather than fundamental entities of the generation layer. The next chapter turns to another concept regarded as fundamental in traditional physics: interaction. It will be shown not to be "force," but the mutual rewriting of generative constraints among frozen structures.

17. Interaction Is Not Force: Generative Constraints Among Frozen Structures

Chapter 16 located particles, mass, and stable structures at the readout layer. This chapter addresses another concept regarded as fundamental in traditional physics: interaction. In physics, interaction is almost always understood as "force"—an action exerted by one object upon another. This chapter aims to show that this understanding must be removed (17.1): interaction is not force, but the mutual prehension of constraints among frozen structures (17.2); it necessarily changes mass, lifetime, and stability, because all three are readouts of the freezing budget (17.3); it is not a local event occurring in spacetime, but a change in freezing reachability (17.4); traditional forms such as "action" and "coupling constants" remain effective because they calculate the statistical geometry of generative constraints (17.5); and force, potential, and field can be recovered as readout projections rather than discarded (17.6).

17.1. Why the Language of "Force" Must Be Removed

The concept of "force" implies four presuppositions, all of which fail at the generation layer. First, it presupposes already existing objects, because force acts upon objects. Second, it presupposes the space in which objects are located, because force has direction and a point of application. Third, it presupposes continuous time, because force causes acceleration, and acceleration is the second derivative with respect to time. Fourth, it presupposes causal pushing, namely "action," because force is the cause of a change of state. These four presuppositions—objects, space, time, and pushing—are precisely the container structures systematically suspended in the preceding chapters. Therefore, "force" is not a neutral concept that can be retained, but the concentrated expression of container grammar in the problem of interaction.

One might wish to retain "force" as an effective approximation. But this does not work within this book, for the following reason: treating "force" as an effective approximation implicitly assumes that force is real above the generation layer and merely approximated. The position of this book is instead that there is no such thing as "force" at the generation layer—force is not an approximated reality, but an interpretive appearance at the readout layer (17.6). The correct approach is therefore not to retain force as an approximation, but to remove it as ontology and then recover it as an appearance at the readout layer. If interaction is not force, what can it be, and where does it occur? The answer is: interaction does not occur in time, but holds in feasibility space. It is not one object pushing another at a certain moment, but a change in the feasibility structure determining "which generations remain viable" among frozen structures. Removing "force" is therefore not a simplification, but a release of structural explanatory power: only when interaction is no longer forced into the framework of "pushing" can its genuine generative content—the mutual rewriting of constraints—be revealed. The conclusion of this section is: the language of "force" must be removed because its fourfold presuppositions are all suspended container structures; interaction does not occur in time, but holds in feasibility space.

17.2. The Minimal Generative Definition of Interaction: Mutual Prehension of Constraints

After removing "force," this section gives a positive definition of interaction. Interaction is not something that occurs "between two objects," because there are no objects at the generation layer. It occurs among frozen structures, and frozen structures are constraints (Chapter 13). Accordingly, the minimal generative definition of interaction is: one frozen structure is incorporated into the constraint set of another generative process—that is, the mutual prehension of constraints. When frozen structure A is prehended as a constraint by generative process B, the set of closeable paths available to B changes accordingly. If B, or its frozen result, is simultaneously prehended by generation associated with A, mutual prehension is formed. This is the entire generative content of interaction.

Three key negations can thereby be clarified. Interaction is not "action": action presupposes an agent and a patient and presupposes causal pushing, whereas constraint prehension contains neither an acting side nor a receiving side, but only mutual changes in constraint sets. It is a symmetric rewriting of qualification rather than one-way pushing. Interaction is not "transmission": transmission presupposes that something is sent from one place to another, whereas constraint prehension transmits no entity, but only changes "which generations are viable." Nothing is sent; only feasibility is rewritten. This is consistent with Section 8.2, where "phase coupling transmits no entity and requires no propagation speed." Interaction is not "exchange": exchange presupposes conserved particles that can be exchanged, such as gauge bosons, whereas at the generation layer, so-called "exchange particles" are readout-layer projections of constraint prehension rather than real exchanges at the generation layer. Three generative consequences of interaction follow: it changes the set of closeable paths, changes the distribution of freezing budgets, and changes stability windows. These three consequences provide the basis for the next section, "why interaction necessarily changes mass, lifetime, and stability." This definition is "minimal" because it introduces nothing beyond constraint prehension: no force, no exchange particle, and no propagation. It is the simplest characterization of interaction using the already established concepts of freezing and constraint (Chapter 13). The conclusion of this section is: interaction is the mutual prehension of constraints among frozen structures. It is not action, not transmission, and not exchange, but the mutual rewriting of feasibility structures.

17.3. Why Interaction Necessarily Changes Mass, Lifetime, and Stability

This section provides a consequence that requires additional mechanisms in traditional frameworks but follows by definition in this book: interaction necessarily changes mass, lifetime, and stability.

A change in mass is not an "effect," but a definitional consequence. According to Chapter 16, mass is a readout of the freezing budget; according to 17.2, interaction changes the distribution of freezing budgets. Therefore, interaction necessarily changes mass. This is not an effect "incidentally produced" by interaction, but a direct logical consequence of the definitions of "interaction" and "mass." Likewise, lifetime must change with interaction: lifetime corresponds to the defreezing threshold and stability of a frozen structure (Section 16.1), while interaction changes the stability window. Therefore, when a structure prehends new constraints, its defreezing threshold changes, and its lifetime changes accordingly. Stability is not "destroyed," but redefined: interaction does not externally "break" a stable structure, but changes the conditions defining "stability" itself. Under a new constraint set, the same structure may no longer satisfy its previous stability conditions. A strong consequence follows: "interaction that does not change mass" does not exist at the generation layer. Any genuine constraint prehension changes the distribution of the freezing budget and must therefore change the mass readout. If a purported "interaction" changes no mass whatsoever, then it is not constraint prehension and therefore not interaction. Mass, lifetime, and stability thus share one generative origin: all are readouts of the freezing budget and stability window, while interaction is precisely the rewriting of the freezing budget and stability window. They are changed not by three different mechanisms, but by the same constraint prehension. The conclusion of this section is: replacing "the action of force" with "the rearrangement of generative feasibility," interaction necessarily changes mass, lifetime, and stability, because all three are readouts of the freezing budgets and stability windows rewritten by interaction.

17.4. Interaction Is Not a Local Event, but a Change in Freezing Reachability

This section addresses the "location" of interaction. The language of a "local event" is invalid at the generation layer because it presupposes a spatial position in which the event occurs, whereas space is a result of the readout layer (Chapter 15). To say that interaction "occurs at a point" is to insert readout-layer structure back into the generation layer. The genuine generative object of interaction is freezing reachability: interaction changes the reachability graph determining "which frozen structures can serve as predecessors of a generation and which generative paths remain reachable." Interaction is therefore not something that "occurs," but something that "continues to hold": it is not a one-time event at some moment, but the continuing validity, in subsequent generation, of the change in reachability produced by constraint prehension.

Scattering, deflection, and delay are therefore only readout phenomena. They are the appearances, at the readout layer and unfolded through temporal–spatial sequences, of changes in freezing reachability, rather than real "collisions" or "deflections" at the generation layer. This also provides a natural unification of interaction and "nonlocality": in traditional frameworks, "nonlocal interaction" is regarded as a problem, whereas in this book it is not an independent concept at all. Freezing reachability is not defined by spatial distance, but only by generative dependence and constraint prehension. Therefore, no presupposition gives priority to "near" over "far." When two frozen structures share or overlap constraints, reachability changes immediately without requiring any "propagation process." At the readout layer, this may appear as "instantaneous correlation," whereas at the generation layer it is merely the result of the same frozen structure being prehended at multiple points. This is fully consistent with Section 8.2, where "phase coupling requires no propagation speed," and Section 15.3, where "the causal cone is freezing reachability." Correspondingly, "interaction propagation" is readout-layer language. Questions such as how interaction propagates, at what speed, and whether it is superluminal are repositioned in this book as questions of how the readout layer maps changes in freezing reachability into temporal–spatial sequences. So-called "propagation speed" is not a generative fact, but a slope readout obtained when a freezing-dependency chain is unfolded into a continuous time parameter under a particular readout protocol. Propagation is not a generative mechanism, but a mode of readout-layer unfolding. The conclusion of this section is: interaction is not a local event, but a structural rewriting of the freezing-reachability graph; scattering, deflection, and delay are readout-layer projections; nonlocality is not a problem because reachability does not depend on distance; and "propagation" belongs to readout unfolding rather than generative fact. Interaction is thereby repositioned from "what happens in spacetime" to "which structures the world permits to continue freezing."

17.5. Why "Action" and "Coupling Constants" Remain Effective

If interaction is not force, why do the entire traditional forms built around "force"—action, coupling constants, variational principles, and field theory—remain so computationally successful? This section answers that question. What requires clarification is not whether they are effective, because they clearly are, but why they are effective.

A coupling constant is the relative weight among freezing budgets: it is not the "strength of force," but a readout quantity describing the relative importance of different frozen structures in constraint prehension. It measures the extent to which one class of constraint prehension changes closeable paths relative to another, which appears at the readout layer as "coupling strength." Action is a statistical-geometric function of closeable paths: it is not a physical entity that is minimized, but a compact encoding of the generative-layer statistical geometry of how closeable paths are distributed. The path that "extremizes action" corresponds at the readout layer to the path with the highest density of closeable paths. Thus, the variational principle is not teleological, but survivor geometry: the system does not choose a path "in order to" minimize action, because that would be teleology, rejected in Section 9.1. Rather, paths at which the density of closeable paths is highest are statistically most likely to be repeatedly frozen and therefore most likely to be read out. The "extremum" is the statistical-geometric appearance of surviving structures, not a goal being pursued. Field theory is effective but has no ontological priority because it correctly captures the continuous approximation of frozen-structure density and the statistical geometry of closeable paths. It is therefore computationally successful. But its success does not prove that the "field" is ontological, just as the success of thermodynamics does not prove that "temperature" is ontological. A field is a readout of frozen-structure density, while temperature is a readout of freezing statistics. This explains an important phenomenon: an "incorrect explanation" can accompany a "correct calculation." A theory may be ontologically mistaken—treating fields as ontology and force as pushing—yet computationally correct, because what it calculates, namely statistical geometry, is real, even though it is assigned an incorrect ontological status. The conclusion of this section is: action, coupling constants, variational principles, and field theory remain effective because they calculate the statistical geometry of generative constraints. Their success is computational success rather than ontological confirmation—they are effective at the readout layer, but possess no ontological priority.

17.6. The Readout Projection of Interaction: The Appearance of Force, Potential, and Field

The final section recovers the previously removed concept of "force" at the readout layer. The old language must be "recovered" rather than discarded, because it genuinely captures real statistical structures at the readout layer. Discarding it would lose effective encodings of those structures, whereas recovering it while explicitly specifying its readout-layer status preserves the computational tools without reintroducing ontological errors.

Force is a response slope at the readout layer: when the constraint environment of a frozen structure changes, its set of closeable paths changes accordingly. At the readout layer, along some unfolded parameter direction, this change appears as a "response slope," which we call "force." Force is therefore not pushing, but a readout of the feasibility gradient caused by changes in the constraint environment. Potential is an effective representation of the distribution of freezing budgets: different positions, understood as organizations of relational density at the readout layer, correspond to different freezing budgets. The distribution of this budget is encoded at the readout layer as a "potential field." Potential is therefore not a real field, but an effective representation of the distribution of freezing budgets. A field is a continuous approximation of frozen-structure density: under finite-resolution readout, the density organization of a large number of frozen structures appears as a continuous "field," consistent with Section 6.5, where continuity is the statistical appearance of freezing. A field is therefore a continuous readout of frozen-structure density, rather than a prior entity permeating space. The three possess internal consistency: force is the gradient of potential, meaning that response slope is the gradient of budget distribution; potential is an integral representation of field; and field is the continuous approximation of freezing density. This traditional force–potential–field relation holds completely at the readout layer because all three are different readout-layer representations of the same frozen-structure density and budget distribution. Their internal consistency does not arise because they reveal the same ontological entity, but because they read different aspects of the same generative-layer statistical geometry. The conclusion of this section is: force, potential, and field are readout-layer projections of interaction. They can be recovered as effective readout-layer language, but none possesses ontological priority—they are readouts of frozen-structure density and generative constraints, not fundamental entities of the generation layer.

17.7: Summary of This Chapter

Chapter 17 has completely liberated "interaction" from the language of "force" and established its generative definition: interaction is not force, not action, not transmission, and not exchange, but the mutual prehension of constraints among frozen structures—a frozen structure being incorporated into the constraint set of another generative process.

This chapter first removed "force" as ontology: its fourfold presuppositions—objects, space, time, and pushing—are all suspended container structures. Force is therefore not an approximated reality, but an interpretive appearance at the readout layer. On this basis, the chapter gave the minimal generative definition of interaction—mutual prehension of constraints—and clarified that it is not action, transmission, or exchange. A definitional consequence follows: interaction necessarily changes mass, lifetime, and stability, because all three are readouts of the freezing budgets and stability windows rewritten by interaction. There is no "interaction that does not change mass." The chapter then repositioned interaction from a "local event" to a "change in freezing reachability": scattering, deflection, and delay are readout projections; nonlocality poses no problem because reachability does not depend on distance; and "propagation" belongs to readout unfolding rather than generative fact. Finally, the chapter explained why action, coupling constants, variational principles, and field theory remain effective: they calculate the statistical geometry of generative constraints, and their success is computational success rather than ontological confirmation. It also recovered force, potential, and field as readout projections and effective readout-layer language, while explicitly denying them ontological priority.

Interaction has thereby been thoroughly repositioned from "what happens in spacetime" to "which structures the world permits to continue freezing." Only one final and most crucial readout-layer concept remains to be fully addressed in Part IV: probability. The next chapter will provide the complete mechanism and falsifiable criteria for probability as a phenomenon of readout truncation.

18. Probability as a Truncation Phenomenon

Chapter 3 diagnosed the suspended status of probability and foreshadowed that the legitimate position of probability lies at the readout layer, with its full mechanism left for Part IV. This chapter fulfills that commitment. It closes Part IV and positively implements the proposition of "no ontic probability" (Section 0.3.2). This chapter aims to show that probability does not belong to generative ontology; it appears only at the generation–readout interface, produced by the triple truncation of gating, bandwidth, and phase fidelity, and each of these three truncations has its corresponding operator at the generation layer, so that they are structured rather than arbitrary (18.1); probability is not ignorance, but structural information loss (18.2); ontic probability would destroy the capacity of physics to fail (18.3); probability must be adjustable, which yields a set of operational experimental criteria, among which Bell correlations constitute a task that must be honestly marked as unfinished (18.4); probability is therefore not the language of the world, but the afterimage of readout (18.5).

18.1. Gating, Bandwidth, and Phase Fidelity

There is no probability at the generation layer, as has already been established repeatedly: the generation layer contains only determinate phase evolution and closure adjudication; a single event either generates or fails, and there is no ontological expression in which it "generates with some probability" (Chapters 3 and 4). Therefore, probability must appear elsewhere—at the generation–readout interface. This section gives the precise structure of this interface: any finite-resolution readout is constrained by three kinds of truncation, and it is precisely these three truncations that make a determinate generative structure appear as a probability distribution. The key point is that these three truncations are not externally imposed experimental limitations, but each has an operator counterpart at the generation layer.

Gating means that readout is effective only within a finite time window, responds only to signals above a threshold, and completely ignores information outside the window. In the language of this book, gating means that phase can be sampled only within a finite time Δt; if phase evolves rapidly within that window, its details cannot be tracked, and the readout can return only "whether it was triggered," not "how it evolved." Formally, when $\Delta\theta = \omega, \Delta t \gg 2\pi$, the readout side cannot resolve the phase structure and can return only a counting result. The "finite time window" here is not an externally imposed experimental parameter in this book, but has an exact generation-layer counterpart: the temporal granularity of readout is determined by the minimal logical step $1/f_{\max}$, that is, by the maximal rhythm $f_{\max}$ among the participating events (following the discrete light-cone bound in Section 15.2). Gating is unavoidable because generative advance itself is discrete and takes $1/f_{\max}$ as its minimal grain; no readout can be finer than this generative grain. Thus gating is not a matter of "the detector not being good enough," but the necessary projection of generation-layer discreteness at the readout interface. This is the key step that elevates gating from an "experimental limitation" to a "structured truncation."

Bandwidth means that readout is also constrained by spectral bandwidth: a detector can respond only to a certain frequency band; high-frequency or low-frequency components are naturally filtered out, and the multiscale structure at the generation layer is compressed into a single-channel output. This means that readout is not an isomorphic mapping of the generative process, but the result of band-pass filtering. Once the spectrum is cut, phase correlations are destroyed, interference information cannot be fully reconstructed, and only intensity-related statistics can be retained. Likewise, the spectral boundary of readout is set by the maximal rhythm $f_{\max}$ among the participating events: bandwidth limitation is not an arbitrary detector property, but the manifestation, at the readout interface, of the intrinsic upper bound of the generation-layer rhythmic spectrum. No readout can respond to structures above $f_{\max}$, because the generation layer itself does not produce rhythms above $f_{\max}$. Bandwidth truncation is therefore also structured and determined by generation-layer parameters, rather than externally stipulated.

Phase fidelity is the true control parameter of probability. In this book, the strength of probability is not an abstract concept, but is controlled by an operational quantity: phase fidelity characterizes whether a readout system can stably track the same generative phase structure across repeated measurements. When phase fidelity is high, the interference structure is clear, the distribution displays distinct phase features, and the "probability cloud" contracts substantially. When phase fidelity is low, phase is averaged out, cross terms disappear, and the statistical distribution degenerates into intensity superposition. Probability is therefore not a "more fundamental description," but a degraded representation after phase information has been lost. The third kind of truncation—the triggerability of threshold gating—has its trigger threshold set by the freezing criteria κ_f, ε_φ, and ε_d, that is, the set of criteria defining the factuality of closure completion (Sections 8.4 and 12.3): readout responds only to structures that cross this threshold, and this threshold is precisely the same set of thresholds by which the generation layer adjudicates "whether something becomes an acquired fact." The operator counterparts of the three truncations are thereby complete: temporal granularity corresponds to $1/f_{\max}$, the spectral boundary corresponds to $f_{\max}$, and the trigger threshold corresponds to the freezing criteria. None of the three is externally imposed; all are determined by the generation layer's own parameters.

This leads to a question that must be answered directly: why does truncation produce probability, rather than merely noise, bias, or unstructured information loss? After all, many kinds of truncation produce only distortion. The key to the answer lies in the point established above: the three kinds of truncation in this book are not arbitrary, but structured—their parameters, $1/f_{\max}$, $f_{\max}$, and the freezing criteria, are all determined by the generation layer itself, and are therefore stable and repeatable. It is precisely this structured and repeatable truncation that allows probability to appear. The argument is as follows: when a settled frozen structure, whose status is determined prior to readout (Section 8.3), is repeatedly acted upon by a set of truncation conditions fixed by generation-layer parameters, what is discarded is always the same class of information—the spectrum above $f_{\max}$, phase detail finer than $1/f_{\max}$, and differences below the freezing threshold—and what is retained is the profile that remains invariant across repeated readout events under this fixed truncation. A probability distribution is precisely this invariant profile. It is not a measure of "what we do not know," but a measure of "what stably survives under structured truncation."

This makes it possible to explain precisely why $|\psi|^2$ repeatedly appears: under the condition that phase is truncated, meaning that fidelity is insufficient, while intensity information remains stably retained across readout events, the intensity-type quantity $|\psi|^2$ is precisely the invariant profile that remains under structured truncation. Therefore

$$P(x) \propto |\psi(x)|^2$$

not because $|\psi|^2$ is "natural probability," but because, after phase has been structurally truncated, $|\psi|^2$ is the only invariant profile that can still stably survive across readouts. The probability law is therefore not a dynamical law, but a law of readout residue—it describes the surviving invariant under structured truncation, not the intrinsic randomness of the world. The strength of this argument must be marked honestly: the argument does not claim to derive a specific probability law from first principles, nor does it claim that deterministic generation by itself is sufficient to reproduce any particular observed statistical law. Its claim is weaker, but substantive: once readout is non-transparent and its truncation is structured by the generation layer's own parameters, there exists a clear path by which a stable probabilistic appearance can arise without treating probability as an ontological or generative primitive; $|\psi|^2$ is here explained as the typical invariant profile along this path, not strictly derived as the unique result.

This section therefore establishes: probability does not belong to generative ontology and appears only at the generation–readout interface; the three truncations each have their generation-layer operator counterpart, $1/f_{\max}$, $f_{\max}$, and the freezing criteria, and are therefore structured rather than arbitrary; phase fidelity determines the form of the probability distribution; and $|\psi|^2$ is the surviving invariant profile under structured truncation, not a fundamental object. Finally, the dependence of this section's argument on the Accessibility Proposition must be stated explicitly: the argument that "probability is a residue produced by structured truncation acting on a settled structure prior to readout" presupposes that the generation layer is at least partially accessible. Only if the truncation parameters of the generation layer, $f_{\max}$ and the freezing criteria, can be tracked, and changes in the readout protocol can produce calculable statistical drift, does the claim that "probability arises from truncation rather than ontic randomness" become substantive and testable rather than an empty assertion. This chapter is therefore the argumentative support for the Accessibility Proposition in Section 0.3.4, and the Accessibility Proposition and the anti-ontic-probability position of this chapter bear responsibility for one another: in fulfilling accessibility, this chapter also fulfills the principle of no ontic probability. This mutual dependence was stated in Section 0.3.4 and is formally carried out here.

18.2. Probability Is Not Ignorance, but Structural Loss

Interpreting probability as "ignorance"—that is, as reflecting our incomplete knowledge of the system's exact state—is a long-standing position. This section aims to show that, although this interpretation points in a roughly correct direction by placing probability at the epistemic rather than ontological layer, it remains insufficient. The position of this book is more precise: probability is not ignorance, but structural and irreversible information loss.

The hidden premise of the "ignorance interpretation" is that there exists an exact underlying state, and probability is merely our incomplete description of it; in principle, if we knew enough, probability would disappear. The problem with this premise is that it provides no criterion for "when probability appears." It only says "we do not know enough," but does not specify under what structural conditions ignorance becomes probability, and under what conditions it does not. Ignorance is a continuous matter of degree without a clear threshold, and cannot explain why probability appears at measurement rather than everywhere. The fundamental turn of this book is this: what probability corresponds to is not ignorance, but irreversible information loss—specifically, the irreversible loss of phase information under structured truncation. Structural loss has three sources, corresponding to the three truncations: gating loses temporal detail, bandwidth loses spectral correlation, and threshold loses subthreshold differences. All three losses are structured, because they are determined by generation-layer parameters, and irreversible, because the discarded information cannot be reconstructed from the readout side.

Precisely because it is structural loss rather than ignorance, probability cannot be eliminated by simply "improving precision." The ignorance interpretation predicts that the more we know, the smaller probability becomes. Structural loss predicts instead that probability is determined by the truncation structure; only changing the truncation structure changes probability, whereas merely improving precision along one dimension leaves probability unchanged if it does not touch the truncation structure. These are two distinguishable predictions, and Section 18.4 will provide experimental criteria on this basis. The "objectivity" of probability is thereby redefined: probability is objective, but its objectivity does not lie in being the intrinsic randomness of the world, or ontic objectivity; it lies in being a repeatable invariant profile produced when structured truncation acts upon an objective generative structure, or readout objectivity. The same generative structure under the same truncation structure yields the same probability distribution; this repeatability is its objectivity. The conclusion of this section is: probability is not ignorance, meaning epistemic incompleteness, but structural loss, meaning the irreversible loss of phase information under structured truncation. It cannot be eliminated by improving precision, but can only be changed by changing the truncation structure. Its objectivity is readout objectivity, not ontic objectivity.

18.3. Why "Ontic Probability" Destroys the Capacity of Physics to Fail

This section returns to the diagnosis of Chapter 3 and deepens it: ontic probability is not only suspended, but would destroy the capacity of physics to fail. This is the most fundamental reason why this book opposes the ontologization of probability.

"Ontic probability" means treating probability as an intrinsic property of the world—the world itself is random, and probability is not a description of any deeper structure but the final fact. The first and fatal problem of ontic probability is unfalsifiability: once probability is the ontic endpoint, any observational deviation can be attributed to "fluctuation," and any counterexample can be deferred to a "larger sample." Ontic probability is therefore immune to all empirical outcomes—it can explain everything, and therefore predicts nothing in a genuine sense (Section 3.3). Ontic probability is logically isomorphic to the continuous limit: the continuous limit makes theory never fail by saying that one can "always approximate one step further" (Chapter 2), while ontic probability makes probabilistic claims never fail by saying that one can "always take a larger sample." They are two versions of the same failure-absorbing mechanism. More seriously, ontic probability conceals generative failure: in this book, the vast majority of generative attempts fail, and failure does not enter the layer of existence (Section 8.5). Ontic probability redescribes these failures as "events that did not occur with some probability," thereby replacing "never entered existence" with "existed with small probability." The ontological status of generative failure is flattened by probabilistic language.

From this, one can see clearly that "random collapse" is not a physical explanation. To say that a measurement result is produced by "wave-function collapse according to Born probability" explains nothing. It merely labels "why this result occurred" as "random," and "random" here functions as a terminator exempting further inquiry, not as a mechanism. A genuine physical theory must allow "complete failure": there must exist decidable conditions under which the theory's core claim is overturned (Section 0.3.3). Ontic probability cancels precisely this possibility. It restates "possible failure" as a "low-probability event," making complete failure impossible to identify. The probability position of this book is therefore explicitly a version capable of failure: probability is the residue of readout truncation, and this claim can be falsified—if changing the readout truncation structure leaves probability unchanged, the claim fails (Section 18.4). The conclusion of this section is: ontic probability destroys the capacity of physics to fail, because it is immune to all empirical outcomes, is isomorphic to the failure-absorbing mechanism of the continuous limit, and conceals generative failure. This book downgrades probability to the readout layer precisely in order to restore its capacity to fail—only a probability that can fail is a physical probability.

18.4. The Adjustability of Probability and Experimental Criteria

The preceding sections established that probability is the residue of structured truncation. This position has a strong and direct empirical consequence: probability must be adjustable. This section turns that consequence into a set of operational experimental criteria, while honestly marking that Bell correlations remain an unfinished task.

Probability must be sensitive to readout structure: since probability is the residue of truncation, systematically changing the truncation structure—gating, bandwidth, and phase fidelity—must cause the probability distribution to vary systematically and calculably. This is fundamentally different from the "probability postulate" view: the Copenhagen interpretation treats Born probability as a non-adjustable postulate, while ontic probability treats randomness as the final fact. Neither predicts that changing the readout structure will change probability. This book treats probability as adjustable readout output and thereby places it under direct experimental pressure. The minimal testable proposition is therefore: while the generative state remains unchanged, systematically changing the parameters of the readout protocol must produce a calculable drift in the observed statistical distribution; if it remains strictly unchanged, the core claim of this book concerning probability fails.

This proposition is implemented as four direct experimental criteria. First, visibility modulation in interference experiments: in double-slit or multipath interference, if coherence time, detection window, and phase stability are systematically improved, fringe visibility must increase monotonically. If fringe visibility remains unchanged after phase fidelity has been independently confirmed to have improved, then the explanation that "probability arises from phase truncation" fails. Second, the response of counting statistics to gating width: in single-particle counting experiments, changing the gating width Δt should change the discreteness of the counting distribution, especially on time scales close to the phase period. If the counting distribution is completely insensitive to gating width, then probability does not arise from gating truncation. Third, continuous degradation of entanglement correlations under phase decoherence: in Bell-type experiments, actively introducing phase diffusion or destroying the reference phase must continuously degrade correlation strength and, in the limit, return it to the local bound. If the correlation remains unchanged after the phase reference is explicitly destroyed, then the "shared phase reference" explanation fails. Fourth, the effect of spectral cutting on statistical distributions: in wave-packet preparation and detection, artificially limiting spectral bandwidth should cause systematic spreading in spatial or temporal distributions. If the statistical distribution is insensitive to spectral cutting, then the Fourier-dual-closure type of explanation fails.

The third item above gives a falsifiable prediction, but it avoids a sharper and more fundamental theoretical question that this book must raise directly and answer honestly: can this framework reproduce correlations that violate the CHSH inequality, CHSH > 2, up to the quantum bound $2\sqrt{2}$?

This question is sharp because of a serious objection: this framework explains entanglement correlations through an inter-event "shared phase reference," and this picture is structurally similar to a local hidden-variable theory—each event carries its intrinsic phase, a kind of "hidden" variable, and correlations are produced by the shared phase structure. But Bell's theorem has proven that any local hidden-variable theory cannot reproduce the correlation statistics of quantum mechanics; CHSH must be constrained by ≤ 2. Therefore, if the phase-reference mechanism of this framework is essentially local-hidden-variable-like, it cannot violate CHSH and cannot reproduce quantum correlations. This would be a substantive blow to the readout-layer program. This book adopts an honest position on this point, stated in three parts.

First, this book does not claim to have constructed a discrete phase model that violates CHSH. This limitation must be stated explicitly. What the book has provided so far is an ontological explanation for the source of correlations—shared phase reference plus readout truncation—and a falsifiable degradation law, namely the third item above. It does not include a proof of the existence of a specific discrete phase update model that has been shown to produce CHSH > 2. Treating "continuous degradation to the local bound" as having solved the Bell problem would be invalid—it precisely avoids the central issue of how to cross the local bound 2 and reach $2\sqrt{2}$. This book explicitly makes no such claim here. This honesty is consistent with the self-limitation of the $|\psi|^2$ argument in this chapter: if even the strict version of the probability residue does not claim to derive a specific statistical law from first principles, the book should still less pretend to have derived a CHSH violation.

Second, whether this framework is necessarily local-hidden-variable-like remains undecided. Bell's theorem rests on a particular combination of locality and hidden variables. Whether the phase-reference mechanism of this framework falls under those assumptions depends on several structural questions that have not yet been clarified: whether the "shared phase reference" of this framework satisfies locality in Bell's sense—bearing in mind that this framework does not presuppose space or propagation at the generation layer, and that Section 8.2.6, "phase coupling requires no propagation speed," and Section 17.4.5, "natural unification with nonlocality," suggest that its correlations may not be spatially local in the first place—and whether difference-based phase updating (Sections 5.5 and 8.2.6) produces correlations beyond the local bound in many-body entanglement. In other words, "this framework resembles local hidden variables" is a judgment that must be tested rather than assumed. This book neither claims to have escaped Bell's constraints nor accepts that it must be subject to Bell's constraints as already proven. Both remain undecided.

Third, this is a constructive task with explicit criteria of success and failure. According to the discipline of the entire book (Sections 0.3.3 and 15.4), this book explicitly lists "whether the framework can reproduce CHSH violation" as an unfinished constructive task, and gives its feasible path and criterion: construct a minimal discrete phase update toy model, the simplest two-body model, each body carrying intrinsic phase and a tendency of directional extension; evolve it according to the difference-based update rules of Sections 5.5 and 8.2.6; apply readout truncation and compute the correlation; then numerically test whether its CHSH value can exceed 2 and approach $2\sqrt{2}$. This toy numerical demonstration does not require an analytic existence theorem—a reproducible discrete-phase numerical counterexample yielding CHSH > 2 would already be sufficient to show that the framework's phase mechanism is not local-hidden-variable-like, thereby blocking the core objection above. Conversely, if, after excluding any still-unidentified spatial-locality assumption, no such construction can exceed the local bound, then the framework's readout-layer explanation of correlations is substantively weakened and should be revised or abandoned accordingly. This book lists this construction as an unfinished item in Appendix G, rather than filling the gap with narrative here.

Therefore, on the Bell/CHSH issue, the position of this book is: it provides an ontological explanation of correlations and a falsifiable degradation law, but honestly acknowledges that it has not yet constructed a discrete model that violates CHSH, and lists that construction as a constructive task with explicit criteria. This treatment accords with the discipline running throughout the book: it is better to mark unfinished work than to conceal gaps with narrative. To use more narrative to "go around" Bell's theorem would be precisely the grammar of deferral diagnosed and rejected in Part I, and this book does not repeat that grammar here.

Finally, the failure conditions of the probability account must be written explicitly rather than hidden in a footnote: if experiments show that, while the generative state remains unchanged, probability distributions are completely insensitive to gating, bandwidth, and phase fidelity, then the entire explanation of probability in this book should be rejected. This is a genuine failure condition. It is an advantage rather than a risk, because it turns probability for the first time into a physical object that can be knocked down by experiment. A theory of probability that can clearly state "it fails here" has greater physical qualification than an ontic probability immune to everything. The measurable proxies mentioned above—coherence time, visibility, the CHSH parameter, and their order-of-magnitude thresholds—belong to the same line of argument as the fourth item of "How to Falsify This Theory" in Part Zero, the readout drift of Section 14.5, and the Accessibility Proposition. Their prediction-card form appears in Appendix H, and the Bell construction task appears in Appendix G.

18.5. Summary of This Chapter: Probability Is Not the Language of the World, but the Afterimage of Readout

Chapter 18 has completed the final step of Part IV and fulfilled the foreshadowing in Chapter 3 and the no-ontic-probability proposition in Section 0.3.2: it has provided the full mechanism and falsifiable criteria for probability as a readout-truncation phenomenon.

This chapter has established that probability never describes "what has happened"—the generation layer contains only determinate phase evolution and closure. Probability appears only at the generation–readout interface. It is produced by the triple truncation of gating, bandwidth, and phase fidelity, and each of these three truncations has its generation-layer operator counterpart, $1/f_{\max}$, $f_{\max}$, and the freezing criteria, so they are structured rather than arbitrary. $|\psi|^2$ is the only invariant profile that stably survives under this structured truncation, not "natural probability." Probability is not "irreducible uncertainty," but the structural and irreversible loss of phase information. It cannot be eliminated by increasing precision, but can only be changed by changing the truncation structure. Its objectivity is readout objectivity rather than ontic objectivity. The chapter further argued that probability cannot return to the ontological layer: ontic probability destroys the capacity of physics to fail, is immune to all empirical outcomes, is isomorphic to the failure-absorbing mechanism of the continuous limit, and conceals generative failure. The only legitimate position of probability is therefore the readout layer. Most importantly, the chapter turned this position into operational experimental criteria—probability must vary systematically with the readout protocol, or the book fails. On the Bell/CHSH issue, it adopted an honest position: it provides an ontological explanation and a falsifiable degradation law, but explicitly does not claim to have constructed a discrete model that violates CHSH; instead, it lists that construction as a constructive task with explicit criteria (Appendix G).

The final conclusion of this chapter is: probability is not the language of the world, but the afterimage of readout. It is not how the world is random, but how we lossily read out a determinately generated world. Part IV is now complete: time, space, causality, particles, mass, interaction, and probability have all been located at the readout layer, as lossy readouts of generative facts rather than fundamental entities of the generation layer. The next part will implement the abstract framework of the first four parts as a minimal model that can be analyzed and reproduced, and will test all the criteria of this book upon it.

Part V.

The Minimal Model and the Appearance of the World

Introduction

In the preceding four parts, this book completed a necessary but essentially destructive task. It systematically dismantled several intuitive structures long regarded as "properties of the world itself" and relocated them one by one to the readout layer: objects do not exist a priori, but are structures after freezing; interaction is not force, but the mutual prehension of generative constraints; the world is not a whole entity, but a statistical appearance; continuity, evolution, and causality are not generative facts, but stable grammars of readout protocols. These conclusions do not constitute a new "worldview," but complete a clearing of layers—which concepts belong to the generation layer, which can exist legitimately only at the readout layer, and which should not be used at all.

After this clearing, however, a sharper question immediately arises: if so much physical intuition has been shown to be appearance, then what remains at the generation layer that cannot be reduced, replaced, or explained away in retrospect? Part V unfolds around this question. The task of this part is no longer to dismantle illusions, but to confirm the bottom line: after continuity, wholeness, causality, and evolutionary narratives have been stripped away, what minimum conditions must the generation layer satisfy for a world first to hold? Which structures exist not because we "see them this way," but because they must be generated in this way?

To answer this question, abstract criteria alone are not enough. The entire abstract framework of the preceding four parts—events, triadic intrinsic structure, differences, closure, freezing, and readout—must be implemented as a concrete model that can be analyzed, reproduced, and fail, and all the claims of this book must be tested upon it. This is precisely the strategy of Part V: it no longer remains at the level of principles, but constructs a minimal generative structure and asks whether it can truly generate stable existence, whether it can provide falsifiable outputs that connect with experience, and where it fails.

Accordingly, this part unfolds around three progressively deepening goals, implemented in five chapters. The first is construction and testing of the minimal model: Chapter 19 gives the four-event Y-type condensate—the minimal nontrivial generative structure this book can construct that satisfies closure and freezing—and shows that its stability comes from directional closure and relational balance, not external constraints. The second is the model's computable outputs and failures: Chapter 20 gives numerical examples based on this minimal model, establishes the discipline of "not pursuing absolute precision, but pursuing differential robustness," and tests falsifiable outputs such as mass differences. Failed examples and successful examples of the model are preserved equally. The third is the emergence from a single structure to the "world": Chapter 21 shows how the juxtaposition of multiple condensates appears at the readout layer as the statistical appearance of a "world," and why abundance is not probability; Chapter 22 then closes continuity, evolution, and causality as readout grammars, completes the de-ontologization of the world, and gives the hook for effective equations at the readout layer—fulfilling the promise made in the preface "For Skeptical Physicists" concerning whether equations can be connected.

Part V does not attempt to reconstruct a "new continuous picture of the world," nor does it provide an alternative holistic dynamics. On the contrary, it pushes generative theory to its most tense position: it retains only those structures without which existence itself could not hold, and implements them as a minimal model that can be numerically tested, pressured by experiment, and explicitly falsified. In this sense, Part V is neither a summary nor an application, but a confirmation contracted to the limit—after everything that can be explained as "appearance" has been so explained, what remains at the generation layer is concretely displayed through a minimal model, and this is what the book truly seeks to defend.

19. The Minimal Condensate Model (Y-Type)

The preceding four parts established the abstract criteria of generation–closure–freezing. This chapter implements them as a minimal model that can be analyzed, reproduced, and fail: the four-event Y-type condensate. It asks: why must the minimal stable structure require at least four events, and why must it be Y-type (19.1); what do directional closure and relational balance in the Y-type mean—and here the relation-only argument is welded in once and for all, blocking geometric circularity (19.2); how phase tension arises within directional closure and provides internal degrees of freedom (19.3); how tension compensation brings about structural stability (19.4); and in which cases this minimal model fails (19.5). The aim of this chapter is not to claim that the Y-type is some specific particle, but to show that the abstract criteria of this book can be concretely satisfied, computed, and tested in a minimal structure.

19.1. Why Four Events

The criterion for a "minimal model" is not the smallest number of events, but the smallest structure that can simultaneously satisfy, for the first time, the three conditions of closure, freezing, and the possibility of failure. On this basis, the cases can be examined one by one. One event and two events are the lower bound of generative meaning: a single event cannot form any relational closure, and two events can couple, but cannot form nontrivial directional closure, since there is only one relation and the directional constraint is trivial.

Three events: closure first appears, since a phase loop can form, but stability degenerates—the directional configuration of three events is too constrained, arbitrary perturbations are difficult to absorb, and freezing cannot be robust to perturbation. Four events: freezing first becomes an adjudicable condition—four events allow the functional differentiation of one central event and three peripheral events; directional closure can first become robust to perturbation, and the freezing window first opens significantly. Therefore, four is the minimal event number that allows "freezing to become an adjudicable condition."

But not any four events will do; the structure must be Y-type. Section 19.2 will argue that, among four-event systems, only the Y-type configuration of "one center and three peripheries" can realize freezeable directional closure. It must be emphasized that the Y-type is not a "fitting result," but a residue of failure filtering: it is not chosen in order to match some known particle; rather, among a large number of four-event configurations, the vast majority are eliminated by directional-closure failure or freezing failure, and the Y-type is the configuration that remains capable of stable freezing. This is consistent with Chapter 9, where "structure is the residue of repeated non-failure." This also defines the responsibility boundary of the minimal model: this chapter does not claim that the Y-type "is" a proton or any specific hadron. It claims only that it is the minimal nontrivial generative structure in which the abstract criteria of this book can be implemented. Its connection, if any, with specific hadrons is the work of Appendices C, E, and F, and it bears responsibility through differential outputs, namely mass differences, rather than absolute matching. The conclusion of this section is: four events are the minimal event number at which freezing first becomes adjudicable; the Y-type is the configuration among them that can stably freeze; and it is the residue of failure filtering rather than a product of fitting.

19.2. Directional Closure and Relational Balance of the Y-Type

The preceding section answered why the minimal stable structure requires at least four events. This section further answers a more specific and stricter question: in a four-event system, why is only the Y-type configuration capable of realizing freezeable directional closure? The conclusion will remain restrained but clear: the Y-type is not an "elegant choice," but the unique nondegenerate solution after directional constraints have been minimized under the condition that event functional differentiation is allowed.

First, it must be clarified that directional closure is not "shape," but the disappearance of constraint. In generative logic, "geometry" does not mean a shape embedded in some background space, but whether directional relations leave an ineliminable residue in the process of closure. Therefore, directional closure means that all directional constraints cancel one another at closure, that no residual directional bias remains, and that the structure can be maintained without external correction. Directional closure is not something "drawn," but the result of the fact that directional constraints vanish.

This point must be stated precisely and positively in order to remove the suspicion of geometric circularity. The following discussion will use expressions such as "directional-gradient," "vector sum equals zero," and "120° configuration," but all of these are bookkeeping representations of directional differences and presuppose no external space. Their exact meaning follows what was established in Sections 5.3, 5.5, and 9.3: direction is the relational extension tendency of rhythm, not spatial orientation, and what enters any adjudication is directional difference—the relative orientation between two extension tendencies—rather than the absolute direction of any event. On this basis, three welding statements must be made explicit. First, what this section calls a "directional-gradient residue" refers to whether the relative orientations among the events' relational extension tendencies resonate with one another or leave an unbalanced net bias, read by the directional kernel k_d, and not to any geometric gradient embedded in space. Second, what this section calls "the sum of the three peripheral direction vectors is zero" and "120°" is a coordinate bookkeeping representation of the relational fact that the relative orientations among the three relational extension tendencies are mutually balanced. This bookkeeping can be rotated as a whole without changing any directional difference, any k_d, or any closure or freezing adjudication, because the generation layer reads only directional differences and the overall rotation of absolute orientations cancels automatically in the differences (Section 5.5). Therefore, "120°" is not the angular position occupied by events in some space, but an equivalent expression of the relational condition that "the relative orientations of the three extension tendencies are balanced." Third, the purely relation-only form of this relational balance condition—using only pairwise directional differences and no absolute coordinates—together with its equivalent expression and verification, is given in Appendix B. Coordinates are used in this section only for intuitive exposition; their ontological content is entirely relational. Accordingly, the directional-closure argument of this section does not presuppose a geometric container and does not constitute a circle of "using geometry to explain the nonexistence of geometry": what is balanced is the relative orientation among relational extension tendencies, what is canceled is the net bias of directional differences, and no external space is required at any point.

With this welding in place, the argument can proceed. In the minimal four-event model, each event carries a directional variable, that is, the relational extension tendency of its rhythm (Section 5.3), indicating toward what kind of relation the event intrinsically extends. The system therefore contains at least four directional extension tendencies, multiple mutually coupled relational chains, and a global directional-compatibility problem. The question is not "how to arrange them more symmetrically," but whether there exists a configuration in which the net directional-difference bias among the extension tendencies is zero at closure. This asks whether the relative orientations among relational extension tendencies can mutually balance, not how events are arranged in space.

Center–periphery differentiation is necessary. If the four events are completely equivalent in function, instability immediately appears: any minute directional bias will be amplified, there is no reference event bearing the coordinating role, and directional closure can depend only on fine-tuning. Therefore, in a four-event system, one functional differentiation must occur: one event bears the reference role for direction and phase, while the remaining events configure themselves directionally around this reference. This is not an artificially introduced hierarchy, but the direct result of minimizing directional constraints. Once center–periphery differentiation holds, the problem becomes: how should the relational extension tendencies of the three peripheral events be configured so that the net directional-difference bias at closure is zero? This is a purely generative problem and is independent of any specific spatial dimension. Its minimal solution is unique: the three extension tendencies are equivalent to one another, the relative orientations between each pair, meaning the directional differences, are equal, and their net biases cancel. In coordinate bookkeeping, this relational balance condition can be written as "the three direction vectors have equal pairwise angles and sum to zero," namely the commonly called 120° balanced configuration. But it must be repeated, following the welding statement above, that this "120°" is a bookkeeping representation of the relational fact that the relative orientations among the three relational extension tendencies are balanced; it can be rotated as a whole without changing any adjudication, and it is not the angular position of events in space. Any deviation from this relational balance will leave residual directional-difference bias that cannot be eliminated, introduce systematic bias into weight distribution, and raise the freezing threshold as a whole.

A common question is why a more "symmetric" configuration, such as one in which all four extension tendencies are completely equivalent, should not be used; in coordinate bookkeeping this would resemble a regular tetrahedron. The reason is not complicated, but is often overlooked: a completely equivalent configuration negates center–periphery differentiation and thereby produces a conflict between directional closure and phase coordination. In generative logic, excessive directional symmetry instead undermines the freezing condition. The Y-type is not the configuration of maximal symmetry, but the lowest and sufficient symmetry under the condition that functional differentiation is allowed. Here too, "regular tetrahedron" and "symmetry" are bookkeeping descriptions of relational configuration and do not presuppose three-dimensional space. They describe the degree of equivalence among the relative orientations of extension tendencies, not a spatial embedding.

The relation between directional closure and the freezing threshold is now clear: directional closure is not itself equivalent to freezing, but it directly determines whether freezing is reachable. If directional closure fails, meaning that the net directional-difference bias is nonzero, the freezing threshold is raised; if directional-difference residue exists, the structure is highly sensitive to perturbation; if relational balance holds, the freezing window expands significantly. The importance of the Y-type configuration therefore lies not in geometric elegance, but in being the first configuration in which the freezing criterion becomes insensitive to directional perturbations. Its mechanistic basis must be added here: directional closure can provide a stable basis for freezing because of the fixity of directional tendencies (Sections 5.3.7 and 13.1). Once the Y-type freezes with a balanced configuration of relational extension, its stability is self-maintained by the fixity of the directional tendencies of the events, rather than by continuous dynamical adjustment. This is also one structural source of the irreversibility of Y-type freezing. But a key limitation must be emphasized: directional closure does not guarantee existence. A Y-type may satisfy directional closure, but this does not guarantee phase closure, nor does it guarantee sufficient weight reinforcement. Directional closure is a necessary condition, not a sufficient condition. This is precisely the position this book has consistently maintained: any structure, even if perfectly balanced in directional relation, may still fail dynamically.

This section therefore reaches a strict, non-rhetorical conclusion: the Y-type configuration becomes the minimal stable structure not because its symmetry is "elegant," but because, under the condition that event differentiation is allowed, it realizes the minimization of net directional-difference bias and relational balance. It is the lowest-order solution of directional balance and the relational precondition under which the freezing criterion first acquires practical meaning. In addition, this section has clarified a methodological point crucial to the whole book: all directional-configuration expressions of the Y-type, including "120°" and "the vector sum equals zero," are bookkeeping representations of the relational fact that the relative orientations among relational extension tendencies are balanced. They presuppose no external space, and their purely relation-only equivalent form is given in Appendix B. Therefore, the directional-closure argument of the minimal model does not constitute geometric circularity—it operates at the level of directional differences throughout, without requiring any geometric container.

19.3. Phase Tension and the Appearance of Internal Degrees of Freedom

Directional closure provides the relational skeleton of the Y-type, but closure is not uniformity. This section aims to show that even when directional closure holds, phase tension remains inside the Y-type, and it is precisely this tension that provides the structure's internal degrees of freedom—making the Y-type not merely a balanced skeleton, but an entity with internal differences capable of carrying structure.

Closure is not uniformity: directional closure ensures that the net directional-difference bias is zero, but it does not guarantee that the phases of the events are identical. In a directionally balanced Y-type, phase differences may still exist between the center and the peripheries, and among the peripheries themselves. Phase is the only degree of freedom capable of carrying tension: once direction has closed and rhythm is compatible, the only remaining carrier of internal difference in the system is phase. Directional differences have been balanced out, rhythmic differences are constrained by compatibility, and only phase differences can exist stably within the closed structure. Accordingly, phase tension is defined as the internal distribution, in a directionally closed Y-type, formed by phase differences among the events and unable to be further erased. Tension is not automatically smoothed out, because erasing it would require breaking directional closure or rhythmic compatibility, which would cause the structure to exit freezing. Phase tension is therefore stably retained as an internal degree of freedom. The minimal definition of an internal degree of freedom follows from this: it is an internal difference that a frozen structure can still carry without breaking closure. For the Y-type, this is the distribution of phase tension.

Phase tension introduces nonequivalence among events: even if the three peripheral events are directionally equivalent, their phases may differ, and they may thereby differentiate functionally. This nonequivalence is precisely the generation-layer entry point for later structures, such as charge, spin, and other readout quantities. The conclusion of this section is: directional closure is not uniformity; phase is the only internal degree of freedom capable of carrying tension; phase tension is stably retained as internal difference in a frozen structure and becomes the entry point for subsequent structural differentiation. This phase tension is precisely the source, in the mechanism for the proton–neutron mass difference in Chapter 16, of the statement that "the neutron's two peripheral events with $\theta = \pi$ introduce stronger phase tension." The greater the tension, the higher the cost of maintaining synchronization, and the larger the correction to the central freezing frequency.

19.4. Tension Compensation and the Appearance of Structural Stability

Phase tension provides internal degrees of freedom, but if tension is not compensated, closure cannot be sustained. This section aims to show that stability comes from the pairing of tension and compensation, and that this pairing is the generative basis of the Y-type as a stable structure.

Without compensation, tension makes closure unsustainable: uncompensated phase tension accumulates and amplifies in subsequent generation, eventually exceeding the freezing window and causing the structure to disintegrate. The basic form of compensation is the redistribution of the generative budget: the system adjusts the distribution of freezing budgets among the events (Section 16.2), allowing tension to be absorbed rather than accumulated. Certain events bear more budget to maintain synchronization, thereby compensating tension. Compensation is necessarily asymmetric: because of center–periphery differentiation (Section 19.2), the central event and peripheral events play different roles in compensation, and compensation cannot be symmetrically distributed. This asymmetry is the source of the internal richness of the Y-type. Compensation is not arbitrary, but constrained by the freezing threshold: only compensation configurations that keep the whole within the freezing window are permitted, while compensation beyond the window leads to defreezing. The tension–compensation pair therefore becomes the source of stable difference: a stable Y-type is not a structure without tension, but a structure in which tension is properly compensated and internal difference is thereby stably retained. Once compensation freezes, it is irreversible: as part of the frozen structure, the compensation configuration derives its irreversibility likewise from the fixity of directional tendencies and the monotonicity of freezing (Sections 11.4 and 13.1). Accordingly, the generative definition of stability is: a state in which a structure can be repeatedly confirmed within the freezing window under a tension–compensation pair. Stability is not absence of tension, but continuous compensation of tension. The conclusion of this section is: tension requires compensation; compensation is the asymmetric redistribution of the generative budget; the tension–compensation pair is the source of stable difference; and stability is the state in which tension is continuously compensated and repeatedly confirmable within the freezing window.

19.5. Failure Modes and the Boundary of the Minimal Model

The final section gives the failure modes of the Y-type model. This is not a defect list for the model, but a positive display of the model's capacity to fail: only a minimal model that can explicitly fail is a genuinely testable model, consistent with Section 0.3.3.

Failure is not an exception, but part of generative logic (Sections 8.5 and 9.5). The Y-type model has four clearly adjudicable failure modes. The first is geometric-closure failure, meaning that the net directional-difference bias cannot be eliminated: when the directional configuration deviates from relational balance, the net directional-difference bias is nonzero, directional closure fails, and the structure cannot enter freezing. The second is phase tension exceeding the compensation window: when phase tension is too large and exceeds what the generative budget can compensate, tension accumulates and amplifies, and the structure disintegrates. The third is competition among compensation paths and locking conflict: when multiple mutually competing compensation paths exist and conflict with one another so that they cannot be simultaneously satisfied, compensation fails and the structure cannot stabilize. The fourth is structural collapse caused by premature freezing: when freezing is triggered too early, before tension has been sufficiently compensated, it locks in an unstable configuration and leads to subsequent collapse.

Together, these four kinds of failure define the applicability boundary of the minimal model: the Y-type can exist stably only when directional closure holds, phase tension lies within the compensation window, compensation paths do not conflict, and the timing of freezing is appropriate. Outside these conditions, it fails. The explicitness of this applicability boundary is precisely an expression of failure responsibility and theoretical honesty. This book does not claim that the Y-type "always" stabilizes, but explicitly states under what conditions it fails, and those failure conditions are adjudicable and numerically testable (Chapter 20). The conclusion of this section is: the Y-type has four clearly adjudicable failure modes, and they define the applicability boundary of the minimal model. Explicitly marking these failure conditions rather than concealing them is the expression, in the minimal model, of the book's discipline of honesty.

19.6. Summary of This Chapter

Chapter 19 has implemented the abstract criteria of generation–closure–freezing as a minimal model that can be analyzed and reproduced: the four-event Y-type condensate.

This chapter established that four events are the minimal event number at which freezing first becomes an adjudicable condition; the Y-type is the configuration among them that can stably freeze; and it is the residue after a large number of configurations have been filtered by failure, rather than a fitted product designed to match known particles. In the argument for directional closure, the chapter welded in, once and for all, a methodological point crucial to the entire book: all directional-configuration expressions of the Y-type—"directional gradient," "vector sum equals zero," "120°," and "regular tetrahedron"—are coordinate bookkeeping for the relational fact that the relative orientations among relational extension tendencies are balanced. They presuppose no external space, can be globally rotated without changing any directional difference or adjudication, and their purely relation-only equivalent form is given in Appendix B. Therefore, the directional-closure argument of the minimal model does not constitute geometric circularity; it operates entirely at the level of directional differences and requires no geometric container. On this basis, the chapter gave the internal structure of the Y-type: directional closure is not uniformity; phase tension is stably retained as the only internal degree of freedom capable of carrying difference; and through asymmetric tension compensation, it brings about structural stability. Stability is not absence of tension, but the state in which tension is continuously compensated and repeatedly confirmable within the freezing window. Finally, the chapter gave four failure modes of the Y-type—geometric-closure failure, tension exceeding the window, compensation conflict, and premature freezing—which define the applicability boundary of the minimal model and fulfill, through adjudicable failure conditions, the book's discipline of honesty.

The abstract criteria of this book have thereby been fully implemented for the first time in a concrete minimal structure, acquiring a form that can be numerically tested and can fail. The next chapter will provide numerical examples based on this Y-type model, establish the numerical methodology of "pursuing differential robustness rather than absolute precision," and test falsifiable outputs such as mass differences.

20. Numerical Examples: Stability, Settlement, and Differences

Chapter 19 presented the structure of the Y-type minimal model. This chapter implements it as analyzable numerical examples, but its purpose differs fundamentally from that of conventional numerical computation: numerical results are not used to prove that the theory is correct, but to expose the conditions under which it succeeds and fails. This chapter will establish why absolute precision is not pursued, whereas differential robustness is (20.1); the principles of minimal specification for the numerical model (20.2); why the numerical criterion of stability is not energy minimization (20.3); how the settlement window is numerically identified (20.4); why all generative outputs can hold only in differential form—with the insensitivity of the proton–neutron mass difference to the central anchor serving as the key numerical demonstration (20.5); and why failed examples are retained alongside successful ones (20.6).

20.1. Why This Chapter Does Not Pursue Precision

In existing physics, numerical precision is almost automatically regarded as a sign of theoretical success: the number of decimal places to which a quantity can be calculated, and the number of standard deviations within which it agrees with experiment, are treated as direct measures of theoretical credibility. This default is reasonable within mature theories, but risky at the present stage of the generative framework. Pursuing numerical precision before the ontological starting point has been stabilized is precisely the practice diagnosed in Part I as "using precision to conceal an incorrect starting point." This chapter therefore repositions the role of "numerics": numerical results are not used for fitting, but as probes for detecting whether the generative mechanism is self-consistent, where it succeeds, and where it fails.

Precise fitting carries structural risks at the generative stage: a model with sufficiently many parameters can almost always fit any given numerical value. Therefore, "successful fitting" does not support the mechanism at the generative stage, and may instead conceal its inconsistency. This is precisely why "differences" are more meaningful than "absolute values": absolute values can be adjusted to any target through recalibration, by changing the unit-mapping constant and zero point, and therefore bear almost no responsibility for failure; differences, such as mass differences, are largely independent of calibration and therefore must either be correct or incorrect, without the possibility of escaping through recalibration. This is fully consistent with the Difference Principle in Section 5.5 and the discipline concerning mass in Section 16.2. Accordingly, the criterion for evaluating the numerical examples in this chapter is not "how accurately absolute values are calculated," but "whether differential outputs are robust, reproducible, and fail when they should fail." Not pursuing precision is therefore not a retreat, but a form of responsibility: it refuses to manufacture a false impression of success using recalibratable absolute values and exposes only falsifiable differential outputs to testing.

20.2. Principles of Minimal Specification for the Numerical Model

The "minimality" of a numerical model does not mean "simplicity," but "incapable of further subtraction," consistent with Section 10.5: if any specification is removed, the model can no longer generate, close, or freeze. Accordingly, minimal specification follows three strict principles. First, only quantities necessary for generation are introduced: the triadic intrinsic structure of events, relational weights, and the criteria of closure and freezing; no background structure, such as spacetime coordinates, continuous time, or probability distributions, is introduced. Second, all parameters are generation-layer parameters: no complete mapping to "real-world scales" is introduced, since that belongs to readout-layer calibration; the numerical model operates only in dimensionless or frequency units at the generation layer. Third, symmetry is not presupposed, but tested: the model is not assumed to possess a particular symmetry; instead, one examines whether the generative dynamics spontaneously produces a stable symmetric configuration.

Why is a complete mapping to "real-world scales" not introduced? Because a complete unit mapping, which maps generation-layer quantities into MeV, seconds, meters, and other units, belongs to readout-layer calibration. Introducing it would immediately bring in recalibratable absolute values and thereby obscure the falsifiability of differential outputs. The minimal model retains only generation-layer structure, postponing calibration until the end and using it only to display absolute quantities (Section 16.2). Minimal specification is therefore not intended to achieve simplicity, but functions as a mechanism of self-restraint: it ensures that numerical examples probe the generative mechanism itself rather than the flexibility of calibration. Under minimal specification, the numerical model is therefore a "structural probe"—it probes whether the generative structure can stabilize and where it fails, rather than whether it can fit a target value. The conclusion of this section is: minimal specification means "incapable of further subtraction" rather than "simple"; it follows the three principles of introducing only quantities necessary for generation, using only generation-layer parameters, and testing rather than presupposing symmetry; and its function is to constrain numerical examples to serve as probes of the generative mechanism.

20.3. Numerical Criteria of Stability Rather Than Energy Minimization

Traditional numerical models almost invariably use energy minimization as the criterion of stability: a system tends toward an energy minimum, and the minimum point is identified as the stable state. Energy minimization is not suitable as a generative criterion, for reasons repeatedly given above (Sections 8.1, 12.3, and 17.5): without continuous time, energy flow cannot be defined; without a presupposed state space, extremum comparison has no domain of definition; and "tending toward a minimum" is teleological rather than generative (Section 9.1). Stability must therefore be adjudicated in a "local and operational" manner, rather than through a global extremum.

This book gives three levels of numerical stability criteria, all of which are local, operational, and capable of failure. The first is phase stability: whether generation has "ceased to drift," meaning whether the phase differences among events cease systematic drift within the settlement window and fall within the closure tolerance. The second is relational stability: whether weights enter an attraction interval, meaning whether relational weights cease systematic increase or decrease and enter a structural basin in the sense of Section 9.2. The third is structural stability: whether the configuration resists perturbation, meaning whether it returns to its original configuration after a small perturbation, rather than disintegrating or drifting elsewhere. Together, these three criteria define "stability," and none appeals to an energy extremum. It must be emphasized that stability is not the endpoint, but a precondition for freezing: a stable configuration is merely capable of remaining; it must still pass freezing, namely repeated confirmation as described in Section 12.4, before becoming a fact. The conclusion of this section is: stability is "capacity to remain" within generation, defined by the three local and operational criteria of phase stability, relational stability, and structural stability, rather than by energy minimization; it is a precondition for freezing, not the endpoint of generation.

20.4. Numerical Identification of the Settlement Window

Numerically, "settlement" must correspond to a window rather than an instant. The reason, consistent with Section 7.3, where simultaneity is defined as a settlement window, is that momentary "exact satisfaction" does not constitute fact, because the next step may immediately deviate. Only "repeated satisfaction within a finite window" constitutes confirmable settlement. The numerical meaning of settlement is therefore the transition from "no longer drifting" to "repeatable": the phases and weights of a group of events repeatedly satisfy the closure criterion within a finite generative-step window $\Delta\lambda$, rather than satisfying it accidentally at a single step.

The minimal identification criterion for a settlement window follows: there exists a finite window $\Delta\lambda$ such that, at every step within the window, the phase-closure tolerance is satisfied, weights lie within the attraction interval, and the configuration resists small perturbations; moreover, this satisfaction does not depend on the specific ordering of steps within the window, consistent with Section 7.3, where "the settlement result does not depend on the internal generative order within the window." The settlement window is "minimal rather than arbitrary": a window that is too short, such as a single step, cannot distinguish accidental satisfaction from genuine settlement; the minimum window length is determined by the minimum number of steps required for repeated confirmation, rather than chosen arbitrarily. Settlement must be distinguished from an attractor and from an equilibrium point: an attractor is the asymptotic terminal state of a dynamical system, and an equilibrium point is a balance of forces, whereas settlement is the repeated satisfaction of a generative criterion within a finite window. It is a discrete and falsifiable adjudication rather than an asymptotic or mechanical condition. The types of numerical failure—a window failing to appear, a window appearing but remaining unstable, or a window depending on step order—are themselves informative: they precisely identify where and how generation fails, corresponding to the failure modes in Section 19.5. The conclusion of this section is: the settlement window is the "minimal evidence of completed generation." It corresponds to repeated satisfaction of the closure criterion within a finite window, rather than instantaneous satisfaction or asymptotic convergence; its minimum length is determined by the number of steps required for repeated confirmation; and the types of numerical failure themselves carry information about the boundaries of the mechanism.

20.5. Differences Rather Than Absolute Quantities

The preceding section gave the numerical identification of the settlement window. This section advances a further key conclusion: all numerical outputs originating from settlement can hold only in differential form; any absolute quantity, unless treated through differences, has no generative meaning. This is not a preference concerning numerical technique, but a direct consequence of generative logic.

Absolute quantities have no anchor in the generative framework. In traditional physics, the legitimacy of absolute quantities depends on some background structure: absolute time depends on presupposed temporal labels, absolute energy depends on a reference vacuum, absolute position depends on a coordinate origin, and absolute phase depends on an external clock. But in the generative framework, there is no a priori spacetime, background vacuum, external clock, or presupposable zero point. Therefore, an apparently harmless expression such as "the frequency of an event is f" is incomplete at the generation layer—not because of insufficient precision, but because it implies an ungenerated object of comparison. Settlement compares "before and after generation," rather than "relative to a zero point": the essence of the settlement window is an internal contrast within the generative process, including before the window versus within the window, unstable structure versus repeated structure, and phase drift versus phase closure. All these adjudications are relational. Therefore, at settlement, the only numerical objects that can be legitimately defined are quantities of change—phase changes, weight increments or decrements, frequency adjustments, and structural differences. In other words, settlement can produce evidence that "change has been locked," but cannot produce evidence of "the numerical value itself."

Differences are therefore generative criteria rather than approximation methods. In many theories, differences are understood as numerical approximations under a continuous limit or as computational conveniences. Here, their status is exactly the opposite: differences are the only numerical form capable of avoiding dependence on a background. For example, within a settlement window, one can legitimately discuss $\Delta f_i = f_i^{(\text{after})} - f_i^{(\text{before})}$, but not an "isolated f_i." The reason lies not in measurement conditions, but in the fact that "before/after" is generated, whereas a "zero point" is not. This is fully consistent with the Difference Principle in Section 5.5: the generation layer uses only differences, and absolute quantities are inoperative. Differences also possess a stronger capacity for failure: they require at least two comparable generative stages, require the stable existence of a settlement window, and are highly sensitive to parameter perturbations. If the settlement window disappears, the difference cannot be defined; if the structure is unstable, the difference continues to drift; if the generative rules are inconsistent, the difference is not reproducible across repetitions. Differences therefore inherently bear responsibility for failure, whereas absolute quantities can often evade failure through recalibration. Accordingly, the numerical examples in this chapter implement the principle of "differences first": they report mass differences rather than absolute masses, compensation magnitudes rather than absolute frequencies, phase-closure deviations rather than "final phases," and whether structures persist across windows rather than whether "structures exist."

Anchor-sensitivity test: Δm does not depend on the choice of f_c. The preceding sections argued in principle that differences avoid background dependence and that absolute quantities lack generative anchors. This section provides a concrete numerical demonstration of that principle and directly resolves a common methodological objection: the model sets the energy scale of the central event at $f_c \approx 930$, corresponding to the dominant contribution to absolute baryon mass, and ultimately reads out an absolute proton mass of approximately 938. Does this constitute fitting in which the input is close to the output?

The key to the answer is to test absolute mass and the mass difference ΔM separately, examining their sensitivity to the central anchor f_c. According to the mechanism established in Section 16.2, the proton–neutron mass difference is

$$\Delta M = \frac{h}{c^2}\Big[\underbrace{(f_d - f_u)}_{\text{dominant flavor-difference term, parameter-free}} + \underbrace{\Delta \bar{f}_g}_{\text{estimated central correction}} + O(\sigma_f^2)\Big],$$

where both terms are frequency differences: the flavor-difference term is the difference between peripheral current-quark frequencies, and the central correction term is the central-frequency difference $\Delta \bar{f}_g$, rather than its absolute value f_c. Thus, f_c enters the dominant term of the absolute mass, but does not enter ΔM. The following test can therefore be conducted: take the central anchor f_c to be a series of different values while holding the remaining generative structure unchanged, including the peripheral phase configurations *uud* versus *udd*, the flavor difference, and the phase-tension mechanism. One should then observe that the absolute mass shifts with f_c, while ΔM remains unchanged. The following table gives the numerical results of this test. The flavor-difference term $f_d - f_u = 3.13 \times 10^{20}$ Hz gives 1.29 MeV, the estimated central correction $\Delta \bar{f}_g \approx 0.5 \times 10^{20}$ Hz gives 0.21 MeV, and the total is reported under the original rounding convention as approximately 1.3 MeV. The numerical values were verified by a reproducible program:

Table 20.1.. Comparison of Strong-Interaction Central-Anchor Variation, Absolute Baryon-Mass Response, and Proton–Neutron Mass-Difference Stability

Central anchor f_c (MeV)	Absolute baryon mass M_{baryon} (MeV)	Proton–neutron mass difference ΔM (MeV)
600	≈ 609	≈ 1.3
800	≈ 809	≈ 1.3
930	≈ 938	≈ 1.3
1100	≈ 1109	≈ 1.3
1300	≈ 1309	≈ 1.3
1500	≈ 1509	≈ 1.3

(The absolute mass $M_{\text{baryon}} \approx f_c+$ small peripheral terms; each peripheral current quark contributes approximately several MeV, and approximately 98% of baryon mass comes from the central event, consistent with the QCD view that baryon mass is dominated by gluons.)

The structure in the table is immediately clear: the absolute baryon mass shifts as a whole with f_c, always taking the form of f_c plus a fixed small peripheral term. This demonstrates precisely that absolute mass is a scale product of the central energy anchor. By contrast, the mass difference ΔM remains unchanged at approximately 1.3 MeV for every value of f_c, because ΔM is determined only by the peripheral flavor difference and the central frequency-shift difference, and is independent of the absolute value of the central anchor f_c. In the program implementation, this invariance is exact: the relative change of ΔM with respect to f_c is zero within numerical precision, because f_c does not appear in the expression for ΔM at all.

This table directly transforms the objection that $f_c \approx 930$ constitutes "fitting one number to another" into an argument in favor of the framework. The substance of the objection is that "930 is input and 938 is output, so the result appears fitted." But the sensitivity test shows that replacing the central anchor with 600, 800, 1100, or 1500 changes the absolute mass correspondingly to approximately 609, 809, 1109, or 1509. This shows that the 930 contained in 938 is purely a translational term of the central energy scale, determining the scale of absolute mass rather than a fitted prediction. Meanwhile, the quantity that genuinely bears predictive responsibility, ΔM, is completely insensitive to the central anchor and remains approximately 1.3 MeV. Therefore, $f_c \approx 930$ is not evidence of fitting. On the contrary, the fact that it can be arbitrarily replaced while ΔM remains unchanged demonstrates that the predictive content of the book's mass claim, namely ΔM, is independent of the choice of central anchor. A quantity that can be freely replaced without affecting the prediction is not a fitted parameter, but precisely the display scale of absolute mass. This is the numerical demonstration of the separation between calibration and prediction.

Two points must be marked honestly, in accordance with the discipline of the entire book and the distinction between primary and secondary terms in Section 16.2. First, the main contribution to ΔM, approximately 1.29 MeV, given by the flavor-difference term bears falsifiable responsibility: it contains no free parameters, is determined by the peripheral phase configuration and the flavor-frequency difference, independently agrees with the experimental value of 1.293 MeV, and connects with high-precision measurements of the proton–neutron mass difference (falsification path F1 in Appendix C). Second, the central correction term $\Delta \bar{f}_g$, approximately 0.21 MeV, is an estimated quantity. Its sign is determined by the phase-tension mechanism—the neutron has greater tension, which slightly raises the central frequency and gives a positive contribution of the same sign—but its specific numerical value is estimated, is entered separately into the error budget (Appendix C), and is not used to rescue the dominant term. The crucial point is that the anchor independence of ΔM does not depend on this estimate. Regardless of the fixed value assigned to $\Delta \bar{f}_g$, because it is a central-frequency difference rather than an absolute value, ΔM remains independent of the central anchor f_c. Its falsification condition is as follows: if strict reproduction finds that the main contribution to ΔM depends significantly on the choice of f_c, or remains incompatible with high-precision mass-difference measurements, the mechanism must be revised accordingly. But so long as ΔM is composed of the flavor difference and the frequency-shift difference and does not read the absolute value of the central anchor, the objection that $f_c \approx 930$ constitutes "fitting one number to another" is fully resolved.

It must be emphasized that prioritizing differences does not mean abandoning connection with experiment. On the contrary, almost all real experiments appear in differential form—spectral peak differences, level splittings, phase drifts, interference-fringe shifts, and changes in correlation functions. The generative framework merely moves this fact forward to the ontological layer rather than repairing it afterward at the experimental layer. The proton–neutron mass difference is precisely a typical differential experimental quantity; see falsification path F1 in Appendix C. The core conclusion of this section is: in generative ontology, existence is not a numerical value, but the fact that a set of differences has been locked within a settlement window. Absolute quantities neither generate nor fail; differential quantities are the only numerical traces left by completed generation. Moreover, through the anchor-sensitivity test, this section has provided a concrete numerical demonstration of the principle: the differential output Δm does not depend on the display anchor f_c, whereas the absolute mass shifts with f_c. This both demonstrates that differences avoid background dependence and fully resolves the objection that $f_c = 930$ constitutes "fitting one number to another."

20.6. Failed Examples Are Retained as Well

The final section addresses a practice rare in conventional numerical literature but indispensable in this book: retaining failed examples. Failure is not "calculation error," but "non-generation." A failed numerical example is not a computational mistake, but a genuine generative result in which the mechanism fails to close or freeze under the given parameter configuration. It is as important as a successful example.

Failure must not be "smoothed away" through parameter scanning. In conventional modeling, nonconvergent cases are often made to "work" by scanning parameters, changing initial values, or adding regularization. This practice is prohibited in this book because it uses the flexibility of calibration to conceal failure of the mechanism, consistent with Sections 20.1 and 20.2. Failed examples reveal the "boundary of the mechanism": they identify precisely where and how the generative mechanism fails, corresponding to the four failure modes in Section 19.5. These boundaries are genuine information about the mechanism rather than defects to be eliminated. Failure is therefore the internal manifestation of theoretical falsifiability: only a numerical model capable of producing explicit failed examples is a model capable of failure and therefore of testing, consistent with Section 0.3.3. How are failed examples "retained" rather than "accumulated"? They are retained as records of mechanism boundaries, marked as "failure due to X under this parameter configuration," rather than accumulated as anomalous data requiring explanation. This is methodologically consistent with the principle that failed structures "do not enter the layer of existence and do not accumulate" (Sections 8.5 and 9.5): failure is recorded as boundary information, not treated as a data point awaiting repair. This differs fundamentally from traditional physics literature, which typically reports only successful fits and conceals failed attempts. This book retains failed examples as formal outputs because failure and success are equally necessary results of generative logic. The conclusion of this section is: failed examples are necessary outputs of generative logic; they reveal mechanism boundaries, embody falsifiability, and are retained as boundary records rather than smoothed away as anomalies. Reporting failure rather than concealing it is the expression, in numerical methodology, of this book's discipline of honesty.

20.7. Summary of This Chapter

Chapter 20 implemented the Y-type minimal model of Chapter 19 as analyzable numerical examples and established the central position running throughout its numerical methodology: numerical results are not used to prove the theory correct, but to expose the conditions of success and failure.

This chapter established that pursuing differential robustness rather than absolute precision is a responsibility rather than a retreat, because absolute values can evade failure through recalibration, whereas differential outputs must either be correct or incorrect. The numerical model follows the principle of minimal specification, meaning incapable of further subtraction, and serves as a probe of the generative mechanism rather than a fitting tool. Stability is defined by three local and operational criteria—phase, relation, and structure—rather than by energy minimization. The settlement window corresponds to repeated satisfaction of the closure criterion within a finite window, and the types of numerical failure themselves carry information about mechanism boundaries. The core of this chapter was the establishment of the principle of "differences rather than absolute quantities," together with its key numerical demonstration—the insensitivity of the proton–neutron mass difference ΔM to the central anchor f_c. The expression for ΔM contains only frequency differences, namely the peripheral flavor difference and the central frequency-shift difference, and does not contain the absolute value of f_c. Therefore, when f_c ranges from 600 to 1500, the absolute mass shifts with it, whereas ΔM remains approximately 1.3 MeV. This anchor independence directly transforms the objection that $f_c \approx 930$ constitutes "fitting one number to another" into an argument, demonstrating the separation between calibration, meaning absolute mass, and prediction, meaning ΔM. Finally, this chapter established that failed examples are retained alongside successful examples: failure is "non-generation" rather than "calculation error," reveals the boundary of the mechanism, and cannot be smoothed away by parameter scanning. Reporting failure is the expression, in numerical methodology, of this book's discipline of honesty.

The abstract criteria of this book have thereby not only been implemented in the minimal model of Chapter 19, but also numerically tested, producing differential outputs that can fail and are independent of anchors. The next chapter will elevate the minimal mechanism of generation–closure–freezing to the "world scale" and show how the "world" appears as the statistical appearance of a large number of frozen structures.

21. Multiple Condensates, Statistical Appearance, and the Emergence of the "World"

Chapters 19 and 20 presented the structure and numerical implementation of a single Y-type condensate. But what we experience is not an isolated condensate, but a "world"—a seemingly continuous and unified whole composed of a large number of structures. This chapter elevates the minimal mechanism of generation–closure–freezing to the "world scale" and completes a key work of de-ontologization. It aims to show that the "world" is not the sum of systems and not an entity at the generation layer (21.1); that the juxtaposition of multiple condensates is not interaction (21.2); that statistical appearance arises from readout sampling of a large number of frozen structures (21.3); that abundance is not probability (21.4); and that the minimal definition of the "world" is a product after readout has been completed (21.5).

21.1. The "World" Is Not the Sum of Systems

The intuition that "the world is the sum of all systems" appears harmless, but in fact implies strong premises. It presupposes that there exists a whole container capable of holding all systems, that the systems simultaneously exist within it, and that their sum constitutes a unified "world." This is the expression of container grammar at the largest scale (Chapter 1). At the generation layer, no "whole" exists: generation is local and path-dependent, and there is no global perspective that simultaneously contains all frozen structures, consistent with Section 7.5, where "history cannot be surveyed as a whole." The totality called "all systems" has no domain of definition at the generation layer.

Therefore, the "world" is a statistical appearance rather than a generative entity. When a large number of frozen structures are sampled, classified, and statistically organized by the same readout protocol, they appear as a seemingly unified "world." But this unity comes from synchronization by the readout protocol, not from wholeness at the generation layer. Multiple condensates do not automatically generate a "world": the coexistence of several condensates does not itself constitute a "world." Only when they are sampled by a common readout protocol and organized into a statistical whole does the "world" appear as a readout appearance. The root of the misplacement of worldview lies precisely in treating readout as ontology: the "unified world that we read out" is mistaken for "a unified entity existing at the generation layer." The conclusion of this section is: the world appears rather than exists. It is not a whole entity at the generation layer, but the statistical appearance presented by a large number of frozen structures under a common readout protocol; the generation layer contains only local frozen structures and no object called the "world."

21.2. Juxtaposition Is Not Interaction

The coexistence of multiple condensates is easily misread as their "interaction." This section blocks that misreading: juxtaposition is not interaction. Juxtaposition originates from synchronization by the readout protocol—multiple condensates are sampled simultaneously by the same readout protocol and appear at the readout layer as "coexisting in one world." But this "coexistence" is synchronization of readout, not coupling at the generation layer. The minimum generative condition of interaction is the mutual prehension of constraints (Section 17.2): two frozen structures constitute interaction only when one is incorporated as a constraint into the generative process of the other. Merely being "read out simultaneously" does not satisfy this condition.

The illusion of juxtaposition has three classical sources: being sampled simultaneously, which is misread as "existing simultaneously in one space"; being classified by the same protocol, which is misread as "belonging to the same system"; and displaying statistical correlation at the readout layer, which is misread as "interaction." But multiple condensates operating in parallel do not imply that they are coupled. Parallelism is common sampling at the readout layer; coupling is constraint prehension at the generation layer. The former does not imply the latter. This distinction is crucial to the appearance of the "world": if juxtaposition is misread as interaction, statistical correlations at the readout layer will be mistaken for genuine coupling at the generation layer, thereby re-ontologizing the "world" as an interacting whole system. The conclusion of this section is: interaction must be generated rather than assumed. Juxtaposition is only synchronized sampling by the readout protocol and does not constitute interaction; only mutual constraint prehension at the generation layer is interaction, and "being read out simultaneously" falls far short of satisfying this condition.

21.3. Where Statistical Appearance Comes From

The world appears statistical—in distributions, fluctuations, and correlations. This section aims to show that statistics is not a natural property, but a mode of presentation. Statistics is not the intrinsic dynamics of the world, but the form in which a large number of frozen structures appear when sampled by a finite readout protocol. When the generative details of each frozen structure cannot be tracked individually and can only be sampled and classified in batches, the result necessarily appears in statistical form. Large numbers are not dynamics, but a sampling condition: in this book, the so-called "law of large numbers" is not a dynamical law governing the world, but an expression of the sampling condition that "statistical appearance stabilizes when the sample is sufficiently large." It describes how readout presents the world, not how the world operates.

How do frequency, phase, and direction "disappear" in statistics? When a large number of frozen structures are sampled through finite-bandwidth and finite-resolution readout, the fine-grained frequency, phase, and directional information of each structure is truncated by the triple truncation of Chapter 18, and only their statistical aggregation is retained. The generative details of individuals "disappear" within statistical aggregation, leaving a distribution. This explains why statistical distributions appear "like laws": stable sampling conditions acting upon stable frozen structures yield stable and repeatable statistical distributions, and this stability and repeatability make them appear as governing laws. In fact, however, they are invariant profiles of sampling, consistent with Section 18.1, where $|\psi|^2$ is treated as an invariant profile. Statistical correlation must be distinguished from generative association: statistical correlation belongs to the readout layer and arises in sampling aggregation, whereas generative association belongs to the generation layer and consists in constraint prehension. The former does not imply the latter, consistent with Section 21.2, where juxtaposition is distinguished from interaction. Statistical appearance is irreducible: it cannot be reduced to precise tracking of every individual, because readout is inherently lossy and collective. Statistics is not "temporary ignorance," but the genuine mode in which the world appears under finite readout. The conclusion of this section is: statistics is the mode in which the world appears, rather than the dynamics of the world. It arises from a large number of frozen structures sampled by a finite readout protocol; large numbers are a sampling condition rather than dynamics; and statistical distributions appear "like laws" because they are invariant profiles of stable sampling.

21.4. Abundance Is Not Probability

Once statistical appearance emerges, an almost unavoidable conceptual slide follows: frequency of occurrence is directly interpreted as probability. The task of this section is precisely to block this slide and to give "abundance" a strict position within generative ontology.

Frequent occurrence does not mean "greater possibility." In conventional physical and statistical language, the reasoning chain "appears more frequently $\rightarrow$ has greater probability $\rightarrow$ is more likely" is accepted almost without reflection. Generative ontology rejects the second step. Within the framework of this book, "appearing more frequently" means only that a certain class of frozen structures forms and is retained more readily under given generative rules, is more readily classified as the same type under a unified readout protocol, and occupies a larger counting share in accumulated samples. None of these points involves an ontological definition of "possibility," still less any prior assignment of probability. Accordingly, abundance is strictly limited to a counting quantity: under given generative rules and a given readout protocol, it is the number or proportion of a certain class of frozen structures that are actually recorded. It depends on the reachability of generative paths, the satisfiability of freezing conditions, and the stability of readout classification; but it does not depend on random selection, assumptions of equiprobability, or a prior probability space. Abundance is a consequential quantity, not an assigned quantity.

Why can probability not be defined at the generation layer? To define probability at the generation layer, at least one of the following conditions would be required: an enumerable totality of possible states, a comparable event space, or repeatable "experimental conditions." Generative ontology explicitly denies all three: generation is path-dependent rather than state-enumerative; once generated, an event freezes rather than being repeatedly sampled; and the concept of "the same experiment" itself arises from classification at the readout layer rather than repetition at the generation layer. This is consistent with Section 9.6, where "what has not occurred has no right to be counted," and with Chapter 18. Therefore, discussing probability at the generation layer is a conceptual transgression. A common objection is that, even without prior probability, long-run relative frequency can serve as the definition of probability. This objection likewise fails within generative ontology: relative frequency already presupposes a stable sample space, whereas the stability of the sample space comes from the readout protocol rather than generative structure. Changing gating, bandwidth, or classification rules systematically changes the frequency distribution. This means that the so-called "probability value" is not an attribute of the event, but a function of the readout rules, consistent with the adjustability of probability in Section 18.4.

If a physical interpretation must be given to "abundance," the most appropriate formulation is this: abundance reflects the channel width of a certain class of frozen structures within the generation–closure–freezing chain. Some phase configurations more readily enter the closure window, some geometrical, meaning relational, structures more readily satisfy the freezing threshold, and some generative paths are more robust to perturbation. Such structures are therefore retained more frequently, not because they are "more likely to be selected," but because they are less likely to fail, consistent with Section 9.1, where "reinforcement means remaining rather than being selected." Misreading abundance as probability produces a series of systematic problems: generative failure is interpreted as a "low-probability event" rather than structural unreachability; model defects are concealed as "statistical fluctuations"; and choices of readout protocol are mistaken for natural laws. Historically, this misreading has been one important source of the prolonged suspension of the measurement problem, the overextension of statistical interpretations, and the philosophical difficulty of the "inexplicability of natural constants." Unlike probability, abundance can fail: changing the freezing threshold so that previously stable structures no longer hold, changing phase gating so that certain generative paths cannot be recorded, or changing classification rules so that objects previously treated as "the same type" are no longer classified together will all change or eliminate abundance. Abundance is therefore not a naturally given constant, but a product of the joint operation of generative structure and readout protocol, and must be subject to failure tests at both the experimental and model levels. The conclusion of this section can be compressed into one sentence: abundance describes "what occurred," whereas probability attempts to answer "what could have occurred"; generative ontology recognizes only the former. The conceptual chain of "statistical appearance–abundance–probability" is thereby completely dismantled.

21.5. The Minimal Definition of the World

The final section gives the minimal definition of the "world." A "minimal definition" is necessary because the preceding sections have shown that the "world" is not a generative entity, not the sum of systems, and not an interacting whole. If the term "world" is still to be used, it must be strictly and minimally redefined; otherwise, it will quietly reintroduce the rejected ontological implications.

The minimal definition of the world, stated operationally, is as follows: the world is the statistical whole presented when a set of frozen structures is sampled, classified, and organized under a common and stable readout protocol. This definition contains several indispensable qualifications. The world is not an object at the generation layer: it has no ontological status there, and the generation layer contains only local frozen structures. The world appears as a statistical whole only after readout has been completed. The world is not the bearer of statistical laws: it does not "obey" some governing law; so-called laws are invariant profiles of stable sampling (Section 21.3), rather than rules obeyed by the world. The world possesses no holistic causality: there is no causality in which "the world as a whole" acts as a cause; causality consists only of local freezing dependence (Section 15.3). The stability of the world comes from the stability of frozen structures and the stability of the readout protocol, rather than from the persistence of a unified entity. The world is "stable" because the frozen structures composing it are stable and the protocol sampling them is stable. The boundary of the world is variable: changing the readout protocol, including what is sampled and how it is classified, changes the boundary of the "world." There is no fixed, protocol-independent "world." The conclusion of this section is: the world is a product after readout has been completed. It is a statistical whole formed when a set of frozen structures is organized under a common and stable readout protocol. It has no ontological status at the generation layer, bears no governing laws, possesses no holistic causality, derives its stability from its components and sampling protocol, and has boundaries that vary with the protocol.

21.6. Summary of This Chapter

Chapter 21 elevated the minimal mechanism of generation–closure–freezing to the "world scale" and completed a key work of de-ontologization: the "world" is not an entity at the generation layer, but the statistical appearance presented by a large number of frozen structures under a readout protocol.

This chapter established that the world is not the sum of systems, because the generation layer contains no "whole," and the apparent unity of the world comes from synchronization by the readout protocol. Juxtaposition is not interaction: juxtaposition is only common sampling, whereas interaction requires mutual constraint prehension at the generation layer. Statistical appearance arises when a large number of frozen structures are sampled by finite readout; large numbers are a sampling condition rather than dynamics, and statistical distributions appear "like laws" because they are invariant profiles of stable sampling. Abundance is not probability: abundance is a counting quantity and the channel width of generation, describing "what occurred"; probability attempts to answer "what could have occurred," has no definition at the generation layer, and, whereas abundance can fail, ontic probability cannot. Finally, the chapter gave the minimal definition of the world: the world is a statistical whole formed when a set of frozen structures is organized under a common and stable readout protocol. It has no ontological status at the generation layer, bears no governing laws, possesses no holistic causality, and has boundaries that vary with the readout protocol.

The elevation from a single condensate to the "world" is thereby complete, and this elevation has not reintroduced any ontological whole: the world remains throughout a product after readout has been completed, rather than the stage within which generation unfolds. The next chapter will address why this "world" appears to evolve continuously, complete the final consolidation of continuity, evolution, and causality as readout grammars, and provide the hook for effective equations at the readout layer.

22. The Readout Grammar of Continuity, Evolution, and Causality

Chapter 21 showed that the "world" is the statistical appearance of a large number of frozen structures. Yet this world also displays three deeper features: it appears continuous, it appears to evolve, and it appears to be driven by causality. This chapter—the conclusion of Part V—locates these three seemingly most unquestionable intuitions as organizational grammars by which the readout layer arranges discrete generative results. It aims to show: continuity is not a generative result, but a readout grammar (22.1, 22.2); "evolution" is not an activity at the generation layer, but the narrative language of traceable differences (22.3); the sense of causality is not a generative relation, but a directional restatement of closure sequences (22.4); and the stability of the empirical world is not guaranteed by generation, but is the result of filtering and reproduction (22.5). Finally, this chapter directly answers the question reserved in the preface for skeptical physicists: can this framework actually connect to a continuous effective equation (22.6)?

22.1. Continuity Is Not a Generative Result

The generation layer contains only discrete facts: events, relation updates, closure, and freezing (Parts II and III). It contains no continuous quantities, no continuous trajectories, and no process that is "continuously occurring." Therefore, "continuous change" cannot be directly derived from closure: closure is a discrete adjudication (Chapter 12), and a sequence of closures gives a discrete sequence of facts, not a continuous trajectory. From "a sequence of discrete closures" to "continuous change," additional structures are required—sampling, interpolation, and smoothing—and these structures belong to the readout layer. Continuity therefore has no generative causal force: it drives no generation, is not the cause of any closure, and is only the appearance of discrete generative results after they have been organized at the readout layer.

Why, then, does continuity appear so "natural" in experience? Because our readout itself has finite bandwidth and finite resolution (Chapter 18). At this resolution, dense discrete frozen structures cannot be distinguished one by one and can only appear as continuous. The "naturalness" of continuity does not arise because the world is continuous, but because our readout cannot resolve discreteness. The conclusion of this section is: continuity belongs to readout grammar, not to generative results. The generation layer contains only discrete facts; continuous change cannot be derived from closure; continuity has no generative causal force; and its "naturalness" in experience comes precisely from the finitude of readout resolution.

22.2. Continuity Is the Product of the Readout Protocol

The preceding section established that continuity belongs to the readout layer. This section gives the specific mechanism by which it is produced within the readout protocol. The finite bandwidth of the readout protocol (Chapters 14 and 18) necessarily produces smoothing: any finite-bandwidth sampling cannot retain high-frequency details, and the filtered-out high-frequency components cause discrete and jump-like generative results to appear at the readout end as smooth continuous curves. Smoothness is not a property of the world, but the necessary result of band-pass filtering. The protocol origin of temporal continuity lies in this: readout samples discrete generative steps with finite temporal resolution and cannot distinguish the discrete intervals between adjacent generative steps; it therefore interpolates them into continuous time (consistent with the observable time of temporal depth in Section 15.1). The protocol origin of spatial continuity is analogous: readout samples a discrete relational network with finite spatial resolution and interpolates dense relational nodes into continuous space (consistent with Sections 15.2 and 6.4).

Therefore, a continuous trajectory is a reconstruction, not a record. The continuous trajectory we "observe" is not the record of a truly continuous path, but the interpolation and reconstruction of discrete frozen results by the readout protocol. The world does not provide a continuous trajectory; readout connects discrete points into a line. Why is continuity so stable? Because the readout protocol itself operates at a fixed resolution and takes continuity as its default interpolation rule. As long as the protocol remains unchanged, the continuous appearance is stably reconstructed (consistent with Section 6.5, where continuity is the statistical appearance of freezing). The conclusion of this section is: continuity is a by-product of the readout protocol. Finite bandwidth produces smoothing, finite spatiotemporal resolution produces interpolation, and continuous trajectories are reconstructions rather than records. Their stability comes from the stability of the protocol, not from the continuity of the world.

22.3. The True Meaning of "Evolution"

The world not only appears continuous; it also appears to be "evolving." This section clarifies the true meaning of "evolution." In traditional physics, the implicit structure of "evolution" is this: a continuously existing object changes its state with a continuous time parameter, and this change is governed by an evolution equation. But at the generation layer there are only sequences, not evolution. What the generation layer gives is the anterior–posterior relation between configurations (logical time in Chapter 7), not "a continuously existing object changing in continuous time." The configurations $\mathscr{C}(\lambda)$ and $\mathscr{C}(\lambda+1)$ are two discrete configurations at different generative depths, not "two moments of the same object."

Where, then, does "evolution" come from? It comes from the traceability of differences at the readout layer. When the readout protocol assigns adjacent frozen configurations to "the same object" and interprets the difference between them as "a change in that object's properties," the narrative of "an object evolving over time" becomes possible. Evolution is therefore not a causal chain, but a narrative chain: it is the narrative structure by which the readout layer organizes a discrete sequence of configurations into a story of "continuous change of the same object," rather than a causal advance at the generation layer. This explains why evolution can be successfully described by equations: evolution equations succeed because they correctly capture the statistical regularities of difference sequences at the readout layer, not because they reveal the true dynamics of the generation layer (consistent with Section 17.5 on why field theory remains effective). But the language of evolution has a boundary of failure: at scales where discreteness at the generation layer becomes significant, or where the readout protocol's attribution of "the same object" fails (such as so-called singularities or critical points of phase transitions), the evolutionary narrative fails and the discrete generation layer shows through (consistent with Section 1.5, where singularities are treated as clues). The conclusion of this section is: evolution is the language of traceability. The generation layer contains only configuration sequences; "evolution" is the narrative chain by which the readout layer assigns adjacent configurations to the same object and interprets their differences as changes of properties. It can be described by equations because equations capture statistical regularities in readout difference sequences, not the dynamics of the generation layer.

22.4. The Readout Origin of the Sense of Causality

The third deep intuition is causality. This section gives the readout origin of the sense of causality. At the generation layer, there is no "causal arrow": the generation layer only adjudicates whether closure holds and whether freezing is triggered (Chapters 8 and 12). It gives a partial order of freezing dependence (Section 15.3), not the semantic differentiation of "cause–effect." The generation layer has no causality of the form "A causes B" with directional semantics; it contains only the dependency fact that "the generation of B depends on the frozen A."

Where, then, does the strong sense of causality in experience come from? It has three readout-layer sources. The first is directional ordering: the readout protocol imposes the partial order of freezing dependence as a directional ordering of "cause before effect" (consistent with Section 15.3, where causality is freezing reachability). The second is conditional stability: when the pattern "freezing of type A is stably followed by freezing of type B" repeatedly appears in readout, the readout layer interprets this stable conditional succession as "A causes B." The third is the principle of minimal description: the readout layer tends to organize large numbers of freezing dependencies in the simplest possible way, and "causality" is a highly compressed descriptive language—it summarizes large numbers of dependency relations with a small number of "causal laws." This also explains the replaceability of "causes" and the non-uniqueness of generative paths: the same B may be achieved through multiple different generative paths, that is, multiple different combinations of frozen predecessors. Therefore, "the cause of B" is not unique. This contradicts the causal intuition of a "single determinate cause," but it is the natural consequence of the non-uniqueness of generative paths. The sense of causality also has a boundary of failure: in situations where generative paths are highly parallel, or where freezing dependencies do not form stable patterns, a clear sense of causality fails. The conclusion of this section is: causality is the readout language of closure. The generation layer contains only the partial order of freezing dependence; the sense of causality comes from directional ordering imposed by the readout layer, recognition of conditional stability, and compression through minimal description. The effectiveness of causal language comes from the success of readout, not from the existence of causal laws at the generation layer.

22.5. Why the Empirical World Is Stable

The final feature requiring explanation is this: why is the empirical world so stable? Stability is not the default property of the generation layer. The default at the generation layer is failure (Sections 8.5 and 9.5): the overwhelming majority of generative attempts do not close and do not freeze. Therefore, the stability of the empirical world is not because the generation layer is "already stable"; it requires a special explanation. Stability has three sources. The first is the freezing-selection effect: only successfully frozen structures enter the empirical layer, while failures never enter it (Section 8.5). The empirical world therefore "appears stable" because we read only those structures that are stable. This is a selection effect, not universal steadiness at the generation layer. The second is the fidelity constraint of the readout protocol: the readout protocol itself can stably sample only frozen structures with sufficient fidelity that can be repeatedly identified. Unstable structures cannot be stably classified by the protocol and therefore do not enter the picture of a "stable world." The third is cross-scale compression and the requirement of consistency: readout must compress frozen structures across different scales in a mutually consistent way, and this consistency requirement further filters out structures that are stable across scales.

Stability is therefore not conservation, but reproduction. A stable structure is stable not because some conserved quantity guarantees that it remains unchanged, but because its freezing mode can be repeatedly regenerated and successfully read out (consistent with Section 12.4, where freezing requires repeated confirmation). Stability is dynamic reproduction, not static conservation. Stability also has conditions of failure: when a freezing mode can no longer be repeatedly reproduced, as in structural decay, or when the readout protocol changes and no longer classifies it stably, stability fails. The conclusion of this section is: stability is a filtered and maintained appearance. It comes from the threefold filtering of the freezing-selection effect, readout-fidelity constraint, and cross-scale consistency. Its essence is repeated reproduction of the freezing mode rather than conservation. Unstable generation has not disappeared; rather, it is systematically unable to enter the empirical layer.

22.6. Effective Equations at the Readout Layer: Fulfilling the Promise of "Can It Connect to Equations?"

The preceding sections have located continuity, evolution, and causality as organizational modes by which the readout layer arranges discrete generative results. A natural and necessary question follows—and this is the question reserved in the preface for skeptical physicists: since the continuous appearance of the world belongs to the readout layer, can this framework actually connect to a continuous effective equation? If it cannot at all, skeptical readers will judge that this framework cannot connect to any computable continuous physics; if it claims to directly provide an ontological evolution equation, it would violate this book's discipline of "not introducing continuous dynamics at the generation layer." The answer given in this section lies between the two, and the precision of this position is crucial.

The answer is: yes, but what it connects to is an effective closure at the readout layer, not an ontological equation at the generation layer. When the generation-layer mechanisms of this framework are coarse-grained at the readout layer, their three classes of mechanisms are projected respectively into three kinds of terms in an effective field equation: co-occurrence-triggered generation is projected as a production term, mesoscopic coordination of phase organization is projected as a transport term, and the irreversible settlement of completed closure (exit from the active sector) is projected as a loss term. This yields a lowest-order single-field effective master equation, with the active frequency-density field $\rho(\mathbf{r},t,f)$ as its object:

$$\frac{\partial \rho}{\partial t} = a_k, f^2, \rho^2 + a_D, f^{-1}, \nabla^2 \rho - a_\Gamma, f, \rho + \mathscr{S},$$

where the frequency scalings of the three coefficients, $k(f) \propto f^2$, $D(f) \propto f^{-1}$, and $\Gamma(f) \propto f$, correspond respectively to the three sectors of production, transport, and loss, while a_k, a_D, a_Γ are probabilistic parameters in the sense of readout residue (Chapter 18). They are jointly determined by generation-layer mechanisms and readout-truncation conditions, and cannot be read off from either side alone.

The strength and boundary of this equation must be marked in three points. First, this equation is a lowest-order effective closure within a controlled closure domain, not a uniqueness theorem: it is valid only in regions where the local number of events is sufficiently large (so that continuity is permissible), anisotropy is sufficiently weak (so that scalar transport holds), memory is sufficiently weak (so that a Markovian density equation holds), and synchronization is weak to moderate (so that single-field closure remains meaningful). Second, the equation and its three coefficients all belong to the readout layer, not to the ontological constants of the discrete-event framework. They are probabilistic parameters of a coarse-grained field equation, not primitive quantities at the generation layer. Third, the three scalings themselves are lowest-order assignments that can be refined by higher-order corrections. The loss term $\Gamma \propto f$ is the most robust, because it directly inherits the irreversible exit of the freezing architecture; the production and transport terms are more sensitive to higher-order correlations and mesoscopic projection assumptions. In other words, what this section gives is not "the equation of the framework," but "the simplest effective equation compatible with the framework after coarse-graining in a controlled readout domain."

Why does this both fulfill the promise and preserve the discipline? It fulfills the promise: this framework can indeed connect to a continuous effective equation that is formally complete, computable, and whose three terms all have sources at the generation layer. The "equation" demanded by skeptical readers appears here. It preserves the discipline: this equation is explicitly characterized as a downstream effective closure, not as an ontological evolution equation. The generation layer remains composed only of discrete events, closure, and freezing (Chapters 8–13); the continuous field equation is only the statistical organization of these discrete facts under finite readout, fully consistent with the conclusion of the preceding sections of this chapter that "continuity and evolution are readout grammars." The full derivation of the equation's discrete origin, the projection from three mechanisms to three sectors, the four classes of closure conditions, and the three failure modes—closure failure under strong synchronization, strong anisotropy, and strong memory—is given in Appendix I (Readout-Layer Effective Dynamics: the GOCD → EQT Bridge).

Finally, one point of resonance should be recorded. Under strong anisotropy, the directional kernel k_d is not eliminated by coarse-grained averaging, but is elevated into a tensorial transport coefficient. This confirms the claim of Chapter 5 that direction is not an optional parallel attribute, but an irreducible dimension of rhythm. The argument is this: if direction were truly separable, an attribute added independently of rhythm, then after coarse-graining, that is, after averaging over the directional degrees of freedom, it should be averaged out and disappear from the effective theory. But it does not; instead, when anisotropy strengthens, it is elevated into a non-eliminable tensorial term. A quantity that is not eliminated under coarse-graining but instead becomes elevated cannot be a separable additional attribute. It can only be an intrinsic dimension of the object being described. Therefore, the non-eliminability of the directional dimension in the downstream effective theory is precisely a downstream confirmation of its status at the ontological layer as an intrinsic dimension of rhythm. The ontological layer says that "direction is an irreducible dimension of rhythm"; the downstream effective theory says that "direction is not eliminated under coarse-graining, but becomes tensorial." The two are upstream and downstream aspects of the same structural fact.

22.7. Summary of This Chapter: The Readout Origin of Continuity, Evolution, and Causality

This chapter has systematically argued for one core conclusion: the most basic and seemingly most unquestionable intuitions of the empirical world—continuity, evolution, and causality—do not belong to facts at the generation layer, but are stable organizational modes by which the readout layer arranges generative results.

Within the framework of generative ontology, the generation layer contains only discrete generative facts—events, relation updates, closure, and freezing—and contains no continuous parameters, continuous trajectories, or "ongoing occurrence" of continuous evolution. So-called "continuous change" cannot be derived from generation itself. This chapter has shown, step by step: continuity is not a generative result, but readout grammar. Continuous time, continuous space, and continuous trajectories all depend on sampling, interpolation, and smoothing rules; they are not observed facts, but representational forms introduced so that frozen results can be ordered, computed, and narrated. "Evolution" is not an activity of the world at the generation layer, but the language of traceable differences. At the generation layer, there are only anterior–posterior relations among configurations. Only when the readout protocol assigns adjacent frozen configurations to the same identity and interprets their differences does the narrative of "an object evolving over time" become possible. Evolution is a narrative structure, not a generative causal chain. The sense of causality is not a generative relation, but a directional restatement of closure sequences. The generation layer only adjudicates whether closure holds; it does not provide the semantics of "cause–effect." The sense of causality comes from the readout layer's directional ordering, conditional stability, and descriptive compression. The stability of the empirical world is not guaranteed by generation, but is the result of filtering and reproduction. Stability does not mean "unchanging," but means that frozen structures can be repeatedly generated and repeatedly read out. Unstable generation has not disappeared, but is systematically unable to enter the empirical layer.

More importantly, this chapter directly fulfilled the promise made in the preface to skeptical physicists: this framework can connect to a continuous effective equation that is formally complete, computable, and whose three terms all have sources at the generation layer, $\partial\rho/\partial t = a_K f^2\rho^2 + a_D f^{-1}\nabla^2\rho - a_\Gamma f\rho + \mathscr{S}$. But this equation is explicitly characterized as a downstream effective closure at the readout layer, not an ontological evolution equation at the generation layer. It fulfills the promise that "an equation can be connected," while preserving the discipline of "not introducing continuous dynamics at the generation layer." Its full derivation is given in Appendix I. The fact that the directional kernel becomes tensorial under strong anisotropy and is not averaged away provides downstream effective-theory confirmation for the claim in Chapter 5 that "direction is an irreducible dimension of rhythm."

Thus, this chapter provides a unified clarification: the world does not continuously evolve and advance causally at the generation layer. The world "appears this way" because the readout protocol stably constructs the appearances of continuity, evolution, and causality upon discrete and failure-capable generative results. This does not deny the empirical world, but strictly locates its layer: continuity, evolution, and causality are retained as effective readout languages, but are no longer permitted to overreach and become ontological explanations at the generation layer. Part V is now complete. From this point on, no discussion of "how the world appears" may be allowed to explain in reverse "how the world is generated."

23. Final Chapter The Universe Is Not a Collection of Existents, but a Process of Completion

23.1. Why a Final Chapter Is Still Needed

In a theory centered on the criteria of generation, closure, and freezing, the task of the final chapter is not to provide a larger picture, but to prevent the picture from being misread as ontology. If, at the end, a theory still attempts to summarize "what the universe is," provide an all-encompassing ultimate model, or declare itself to have completed the final explanation of reality, then it has structurally violated the principle this book has insisted on from the beginning: any unfailable holistic narrative is an unauthorized generative assumption. Therefore, this final chapter does not provide a "world picture," introduces no new generative mechanism, and does not attempt to extrapolate the framework of this book into an answer to all questions. It completes only three irreplaceable tasks that are often avoided.

First, it clarifies what this book has actually completed. This book has not proposed a theory of the "contents of the universe," but has completed a structural reconstruction of how existence can hold. More precisely, what this book has completed is this: it has redefined "existence" from an attribute of objects into a completed state of generation; redefined "structure" from geometry or entity into a relational configuration capable of closure and freezing; repositioned the "world" from an ontological whole into a stable appearance at the readout layer; and systematically reduced continuity, evolution, causality, and probability to readout languages rather than generative facts. These conclusions are not explanatory or intuitive, but have been established step by step through a full set of generative criteria that can fail, be blocked, and be bounded.

Second, it clarifies what this book has deliberately not done. Equally important is that this book has deliberately refused to undertake the following tasks: it does not provide a global dynamics of the universe; it does not assume a global state, global time, or global probability space; it does not attempt to unify all physical theories; and it does not extrapolate the generative framework into a direct description of the empirical world. These refusals are not conservatism, but structural discipline—because once these tasks are accepted, generative logic inevitably slides back into totality, continuity, and unfailable narrative. This book would rather remain incomplete than purchase an appearance of completeness through overreaching explanation.

Third, it clarifies where this theory is willing to fail. A theory of generation has no ontological significance if it cannot state where it would fail. This book therefore explicitly assumes the following risks of failure: if the closure condition cannot be clearly operationalized, the framework fails; if the freezing criterion cannot be distinguished from the readout layer, the framework fails; if generative failure cannot be recorded, classified, and retained, the framework fails; if continuity and causality are reintroduced as generative causes, the framework fails. These failures are not defects, but public commitments concerning the scope of applicability of the theory (they are the closure of the same set of commitments expressed in the seven criteria of Part Zero, "How to Falsify This Theory").

Thus, the final chapter is not meant to "end the discussion," but to prevent the discussion from sliding into unauthorized totalizing narrative. Here, this book completes only one thing: it reunderstands the "universe" from an already existing whole into the result of a series of generative processes completed, frozen, and read out at different layers. The universe is not a collection of existents, but a history of completion; and the responsibility of theory is not to describe "what it is," but to state under what conditions it qualifies as "existence."

23.2. The Concept of "Existence" Must Be Rewritten

Across all the preceding chapters of this book, a series of negative conclusions has been systematically completed, and these conclusions have one common consequence: the traditional concept of "existence" has lost its legitimate context. This book has argued and repeatedly verified that there is no a priori spacetime as a container of existence (Chapter 1), no continuous evolution as the mechanism maintaining existence (Chapter 2), no self-sufficient object as the bearer of existence (Chapter 4), and no ontic probability as the principle distributing existence (Chapters 3 and 18). Within such a generative framework, continuing to use the word "existence" without rewriting it is no longer merely linguistically imprecise; it is an overreach across conceptual layers. The first task of this chapter, therefore, is not to introduce a new cosmic picture, but to structurally rewrite the concept of "existence" itself.

The traditional concept of "existence" relies on three undeclared premises. The premise of persistence: an existent is assumed to maintain itself in time, and change is treated as a change of its attributes rather than a rupture of its existence. The premise of location: an existent is assumed to be "located" at some spatiotemporal position, and changes of position do not affect its qualification as existent. The premise of repeatable identification: the same existent can be recognized again as "the same" across different times and under different conditions. Yet within the framework of generative ontology established by this book, all three premises fail: generation is a discrete sequence of events, not a continuous process of maintenance; events are not embedded in spacetime, since spacetime itself is a product of the readout layer; and frozen structures are only transiently stable results in generative dynamics, not entities that can be identified without limit. Therefore, the traditional concept of "existence" is not corrected, but withdrawn as a whole.

This book's replacement for "existence" does not introduce a new entity name, but a strict adjudication condition: to exist is to be the result that has completed closure and been frozen within generative dynamics. This definition has three irreplaceable features. Criteriality: existence is not described into being, but adjudicated; the question is no longer "what is it," but "has it completed." Process-dependence: existence depends neither on background, nor container, nor a priori coordinates; it depends only on whether the generative path has successfully crossed the thresholds of closure and freezing. Capacity for failure: existence is not a necessary result; most generative attempts do not form existence. This is the theory's expected result, not an anomaly. In this sense, "existence" is no longer a noun, but the completed state of a verb.

Rewriting "existence" as "completion" is not a semantic replacement, but an ontological reconstruction, and it brings a series of irreversible consequences: there are no "objects not yet completed but already existent," no "partial existence" or "potential existence," and no qualification of existence independent of the generative process (fully consistent with Section 9.6, where "possibility means generability, and what has not occurred has no right to be counted"). A structure either completes or does not complete; and what has not completed has no ontological status. This means that existence is no longer a basic fact of the world, but the rare result of generative success. If this rewriting is refused, all the work of the preceding five parts will be systematically canceled: continuity will be quietly reintroduced as the way existence is maintained, probability will be reinterpreted as the principle distributing existence, the world will be reunderstood as an already given whole, and failure will once again be interpreted as "not yet occurring." Therefore, this rewriting is not a philosophical declaration of position, but the final guard of the book's logical consistency. The conclusion of this section can be stated precisely in one sentence: in generative ontology, existence is not the starting point, but the result of a generative process successfully completed under strict criteria. From this point onward, the legitimate use of the word "existence" in this book is permanently restricted—it may refer only to structures that have completed closure and been frozen, and may no longer be used as any premise explaining generation.

23.3. The Universe Is Not a "Whole"

After the rejection of self-sufficient objects, an extremely common and deeply concealed theoretical slide appears: the bearer of "existence" is transferred from objects to "the whole." This slide usually appears in a seemingly modest form—it no longer speaks of particles, entities, or ontological units, but turns instead to the "universe as a whole," the "global state," or "total evolution." This book must state clearly: this substitution is equally illegitimate.

In generative ontology, the reason "objects" are rejected is not that they are too small, but that they assume a bearer unit that can be continuously identified. The "universe as a whole" repeats the same error at a higher layer. In generative logic, there is no "set of all events" that can be enumerated simultaneously, no "global configuration" that can be adjudicated for closure all at once, and no global time on which "total evolution" could unfold (consistent with Section 7.5, where "history cannot be surveyed as a whole," and Chapter 21, where "the world is not the sum of systems"). The "universe as a whole" is therefore not a generative object, but the mistaken elevation of the statistical appearance of the readout layer, namely the "world," into an ontological whole at the generation layer. The position of this book is this: the universe is not an already existing entity that can be identified as a whole, but the never-surveyable accumulation of a large number of local generation–closure–freezing processes. It has no "global state," because there is no global perspective capable of simultaneously containing all frozen structures. It has no "total evolution," because evolution itself is narrative language at the readout layer (Section 22.3). To treat the universe as a whole is the return of container grammar at the largest scale and for the final time; this book suspends it here for the final time.

23.4. Time Is Not Passage, but the Sequence of Completion

The third concept that must be finally clarified in the final chapter is time. Traditional intuition understands time as a uniformly flowing background in which things persist, change, and age. This book has systematically rejected that picture: time is not a parameter of passage, but the monotonic trace left after generation has been frozen.

This conclusion is not asserted once in this book, but established across four levels. The final chapter only cites their closure rather than repeating the argument. Its status as a principle appears in Section 11.4: monotonicity, as an institutional principle, stipulates that "rollback after freezing is illegitimate." Its formalization appears in Section 7.4: the frozen set is monotonically nondecreasing with generative steps, $\mathscr{F}(\lambda) \subseteq \mathscr{F}(\lambda+1)$, and the distinction between "same-layer non-reopening and higher-level inheritance" makes this a precise and decidable structural proposition, thereby deriving the arrow of time as the monotonic retention of acquired status. Its mechanistic basis appears in Sections 5.3.7 and 10.2: the fixity of directional tendencies allows frozen structures to maintain themselves, and the asymmetry by which phase updates at each step while direction is realized at each step but does not update is the structural source of irreversibility. Its full development appears in Chapter 13. The final chapter restates only the essential point: the directionality of time requires neither entropy, nor low-entropy initial conditions, nor any statistical assumption. It comes directly from the monotonic accumulation of freezing. What this book establishes is a structural arrow of time—the minimal generative asymmetry produced by the monotonic retention of acquired status. It stands upstream from the thermodynamic arrow and is the more fundamental asymmetry that any later theory of time must respect if it seeks to ground directionality. Therefore, in the final picture of this book, the universe does not unfold along a flowing time axis, but consists of a sequence in which completions are irreversibly frozen and thereby accumulate into the "past." Time is not the stage of the universe, but the trace left by completion.

23.5. Probability Is Not an Attribute of the World

The fourth concept that must be finally clarified is probability. Traditional physics treats probability as an intrinsic attribute of the world—the world itself is random. This book has systematically rejected this position: probability is not an attribute of the world, but the truncation residue of finite readout applied to determinate generative results.

This conclusion is likewise not reargued in the final chapter, but only cited in closure. Its diagnosis appears in Chapter 3: probability has never been defined at the generation layer, appears only after measurement, and varies with the mode of measurement. Its full mechanism appears in Chapter 18: the triple truncation of gating, bandwidth, and phase fidelity each has a generation-layer operator counterpart, $1/f_{\max}$, $f_{\max}$, and the freezing criteria; probability is the surviving invariant profile under this structured truncation, and $|\psi|^2$ is explained here as a typical invariant profile rather than natural probability. The final chapter restates only two essential points. First, probability is therefore capable of failure: this book's core claim about probability can be directly negated by experiment. If, under an unchanged generative state, the readout protocol—gating, bandwidth, and phase fidelity—is systematically changed while the statistical distribution remains strictly unchanged, then the claim that "probability arises from readout truncation" fails (Section 18.4). This is one of the most direct failure criteria of this book. It transforms probability from an ontological stipulation immune to everything into a physical object that can be brought down by experiment. Second, the strength of this claim is exactly equal to the strength of the Accessibility Proposition in Section 0.3.4, and the two bear responsibility for one another. In fulfilling "probability arises from truncation," this chapter also fulfills "the generation layer is at least partially accessible." If accessibility is falsified by the same class of experiments, then the anti-ontic-probability program fails together with it. Therefore, in the final picture of this book, the world does not occur randomly; it is determinately generated and then finitely read out. Probability is not how the world throws dice, but how we lossily read out a generative history that throws no dice.

23.6. The Responsibility of Theory: Failure, Not Explaining Everything

At the end of the final chapter, this book returns to the only posture it has insisted on since the preface: the responsibility of a theory is not to explain everything, but to state where it would fail.

This book has not provided a complete picture of the universe, and explicitly refuses to provide one. It provides a set of criteria—what existence is, how structure holds, how the world appears, and under what conditions all of these fail to hold. It exposes its most central claims to positions where they can be negated: if the closure criterion cannot be operationalized, this book fails; if freezing can be reversed at the same layer, this book fails; if probability does not vary with the readout protocol, this book fails; if background, continuity, or ontic probability must be reintroduced in order to maintain the framework, this book fails. It also honestly lists the thresholds it has not yet crossed—the recovery of Lorentz invariance, the emergence of gauge groups, the discrete regularization of ultraviolet divergences, and the discrete reproduction of Bell correlations (Section 15.4 and Appendix G)—and treats them as constructive tasks with explicit criteria, rather than concealing them as "solved in principle."

This posture is not humble rhetoric, but the logical endpoint of the entire argument of the book. A theory capable of explaining everything loses its physical qualification precisely because it cannot fail. A theory that can specify precisely "it collapses here," whether or not it ultimately holds, has already stood on the side of physics. This book would rather be the latter. If, after serious examination, the reader can clearly identify where this book fails, that is precisely the result this book hopes to elicit—because a theory that cannot be clearly negated is what is truly unphysical.

The universe is not a collection of existents, but a process of completion. This sentence is not a new picture of the world, but a condensation of a criterion: what has not completed has no existence; what has completed cannot be withdrawn; and the world of continuity, evolution, causality, and probability that we experience is the appearance produced after these completions are finitely read out. The entire work of this book has only been to carry this criterion through to the end, while marking at every point where it may be wrong.

24. Afterword Written After the Completion of Generative Logic

24.1. Why This Book Had to Be Written This Way

The fact that this book seems "unlike a standard work of physics" is not a stylistic choice, but is determined by the problem it addresses. When a theory simultaneously rejects a priori spacetime, continuous evolution, ontic probability, and unfailable propositions, it cannot continue to follow the familiar technical narrative path: given a background $\rightarrow$ write down equations $\rightarrow$ fit phenomena. This book does not discuss one kind of interaction, one class of physical effects, or even a model that can be connected directly to experiment. It discusses a more basic question: without background, without objects, and without probability, how is a physical world possible? At this layer, to give "computable results" prematurely would not be efficient, but irresponsible. This book therefore chooses a more difficult path, one also more easily misunderstood: not to calculate first, but first to rewrite generative logic itself.

24.2. Clarifying the Relation to the Previous Nine EQT Works

The nine EQT works completed earlier involved a unified perspective on the four interactions, generative explanations of dark phenomena, measurement and readout mechanisms, process philosophy and the mind–matter problem, and a systematic return to the Planck problem. Those works are neither mistaken nor denied. But they share a fact that must now be explicitly pointed out: they still operate within the technical grammar of the continuous limit and differentiable structures. In those works, spacetime was reinterpreted as gradients and distributions, probability was weakened into a statistical result, and objects were de-substantialized into density structures—but this rewriting still took place within an already given continuous framework. What this book has completed is an earlier cut: it no longer assumes any differentiable structure, any continuous parameter, or any a priori measure. Therefore, this book is not a summary or upgrade of the preceding nine works, but a reset at the generation layer. In this sense, the previous nine works are application-layer works that can be translated back into this framework, while this book is the generative foundation they did not previously possess.

24.3. Why This Book Deliberately Leans Toward Philosophy Rather Than Physics

A common question is: if the goal is a physical theory, why not prioritize experimental testability? The answer is not complicated. When the starting point of a theory itself contains a logical circle, more refined equations only reinforce the error; more complex models only conceal the premises; and fittings closer to experiment only postpone failure. At such a stage, continuing to push for "physical results" is not courage, but avoidance. This book therefore chooses a posture that is not well-liked in contemporary physics: to first assume philosophical responsibility, and only then physical responsibility. Here "philosophy" does not mean rhetoric, metaphor, or a worldview declaration, but the rigorous treatment of the following questions: what can serve as a unit of generation, what can serve as a criterion of completion, and what can count as failure. Only after these questions have been clearly answered can any further physical model have testable meaning.

24.4. How This Book Hopes to Be Read

This book is not suited to the following expectations: looking for new formulas, quickly obtaining experimental predictions, expecting an immediately usable alternative model, or hoping to quickly judge whether it is "right or wrong." Instead, it is suited to physicists willing to temporarily suspend existing theoretical frameworks, researchers uneasy with the fact that "everything always begins from continuity," and readers who remain vigilant toward the concept of "existence." The ideal mode of reading is not linear advance, but repeated return: when the "closure" of Chapter 10 becomes clear, return to Chapter 1; when the "probability truncation" of Chapter 18 seems radical, return to Chapter 3; when the minimal model of Chapter 19 appears too simple, return to the failure principles of the entire book.

24.5. On Misreadings

This book does not expect to be widely accepted, but the following misreadings must be explicitly excluded: this is not an "anti-mathematics" position, not an "anti-experiment" posture, not a retreat in which "physics goes back to philosophy," and not any form of metaphysical manifesto. If one sentence is needed to prevent misreading, it should be this: this book is not answering "what the world is," but asking "how the world is allowed to appear."

24.6. After the Writing Is Complete

Completing this book does not mean that generative logic has been finally completed. It means only one thing: from this point onward, any further work based on this framework should not return to a background-dependent starting point. Whether it continues, expands, fails, or is overturned should all occur within the generative framework delineated here. If future work shows that closure cannot stabilize, freezing cannot form, or readout cannot be unified, then this book should be abandoned. There is no regret in this—because a theory that cannot be abandoned does not deserve to be taken seriously.

24.7. Closing Words

This book was written slowly, not because the calculations were difficult, but because each step had first to dismantle existing foundations. If readers feel discomfort, hesitation, or resistance during the reading, that is not a side effect, but a signal: generative logic is taking effect. Whether the universe is truly like this will be decided by the future. But at least here, a starting point that does not depend on background has been written down in full.

— The End —

A. The GOCD Axiomatic System of Event Sequences (Mathematical Formalization)

This appendix provides a testable, derivable, and falsifiable formal skeleton for Generative Ontology with Closure Dynamics (GOCD). The basic claim of GOCD is that the primordial structure of the universe is not presupposed spacetime or presupposed entities, but a countably discrete event-sequence system; spacetime, matter, interaction, and quantum statistics all emerge from the triadic intrinsic properties of events (frequency, phase, and direction), together with the phase evolution, synchronization, and closure mechanisms driven by relational coupling among them.

To ensure logical closure and satisfy the requirement of falsifiability, this system adopts a layered structure of "axioms–definitions–rules–theorems (or propositions)–falsification conditions," and follows two modeling principles. The first is formal closure: all evolution is given by internal rules and does not depend on any external spacetime background or externally added coordinate system. The second is testability: any newly introduced structure (constants, thresholds, kernel-function forms, projection operators, triggering conditions, and so on) must correspond to comparable observational outputs and explicit falsification paths, and must not serve merely as an explanatory supplement. The structure of this appendix is as follows: basic objects and axioms (A1–A6), the seven diagnostic tests for a genuine settlement predicate, dynamical rules (D1–D5), conservation and monotonic laws (C1–C3), emergence mappings (E1–E4), and falsification conditions (F1–F4).

A.1. Basic Objects and Axioms (A1–A6)

Definition A0 (Cosmic Configuration). Over discrete logical time $\lambda \in \mathbb{N}$, the universe is in the configuration

$$\mathscr{C}(\lambda) = (\mathscr{E}(\lambda),, W(\lambda),, \mathscr{F}(\lambda)),$$

where $\mathscr{E}(\lambda)$ is a countable set of events, $W(\lambda) = [w_{ij}(\lambda)]$ is the relational weight matrix, and $\mathscr{F}(\lambda) \subseteq \mathscr{E}(\lambda)$ is the subset of frozen events.

Axiom A1 (Event-Sequence Axiom). The universe is composed of countable event processes: for each step λ, there exists an event set $\mathscr{E}(\lambda)$, whose evolution is given entirely by internal rules and does not depend on any external spacetime background or externally added coordinate structure.

Axiom A2 (Triadic-Property Axiom). Each event i carries triadic intrinsic properties $(f_i, \theta_i, \mathbf{d}_i)$, where f_i is rhythm (frequency), θ_i is phase, and $\mathbf{d}_i$ is direction (the tendency of relational extension). At this layer, f_i is not required to be dimensionalized a priori; its connection with experimental dimensions is completed through the emergence-mapping module (E1–E4) and is subject to the external constraints of the falsification-condition module (F1–F4).

Axiom A3 (Relational-Weight Axiom). Interaction among events is expressed only through weights $w_{ij}(\lambda) \in [0, 1]$, satisfying the following: (symmetry, minimal form) $w_{ij}(\lambda) = w_{ji}(\lambda)$, which may be relaxed under a directed-causality extension without destroying computability; (locality constraint, sparsity) for any event i, its effective neighborhood is a finite set; (computability) there exists a computable function F such that weight updating is determined by

$\{\Delta f_{ij}, \Delta\theta_{ij}, \mathbf{d}_i \cdot \mathbf{d}_j, \eta\}$, where $\Delta\theta_{ij} = \mathrm{wrap}(\theta_i - \theta_j) \in (-\pi, \pi]$, and η denotes a finite number of model constants/thresholds. The introduction of any η must provide observable output and a falsification path in the F module.

Axiom A4 (Closure–Existence Axiom). A group of events acquires existential status if and only if it satisfies a decidable closure condition and is frozen. This axiom corresponds to the formal version of the book's proposition "completeness is existence": closure is the threshold of existence, and stability is expressed through decidable conditions rather than narrative assertion. The diagnostics it must pass as a "genuine settlement predicate" are given in Section A.2.

Axiom A5 (Generative-Conservation Axiom). The generation and transformation of events satisfy a conserved scalar $\mathscr{S}$ ("frequency share"): in the absence of external injection and when generation uses a conservative redistribution operator, $\mathscr{S}$ is conserved under stepwise evolution.

Axiom A6 (Non-Probabilistic-Ontology Axiom). The ontological evolution layer contains no probabilistic component; probability is written only into the readout layer and sampling layer, not into the ontological evolution layer. This axiom is consistent with Chapter 18, where "probability arises from truncation."

A.2. The Seven Diagnostic Tests for a Genuine Settlement Predicate

Axiom A4 stipulates closure as the threshold of existence. But for "closure" to count as a genuine settlement predicate (Section 8.3), rather than merely "reaching a certain state" or "being retrospectively recognized by later readout as completed," it must pass the following seven diagnostic tests. Let the candidate closure set be $\mathscr{C} \subseteq \mathscr{E}$, its closure criterion be denoted by $\mathrm{Cl}(\mathscr{C})$, and the corresponding settled status be denoted by $\mathrm{Settled}(\mathscr{C})$.

T1 (Non-Circularity). The definition of $\mathrm{Cl}(\mathscr{C})$ may refer only to quantities of the generation layer itself—rhythmic differences Δf_{ij}, phase differences $\Delta\theta_{ij}$, directional differences $\mathbf{d}_i \cdot \mathbf{d}_j$, and the weights and thresholds constructed from them—and must not refer to any readout-layer operator, observational record, or sampling result:

$$\mathrm{Cl}(\mathscr{C}) = \mathrm{Cl}\big(\{\Delta f_{ij}, \Delta\theta_{ij}, \mathbf{d}_i \cdot \mathbf{d}_j\}_{i,j\in\mathscr{C}}\big),$$

with no $\mathscr{R}$ (readout protocol) appearing.

(Corresponding to "internality" in Section 12.2.)

T2 (Substantive Nontriviality). Settled must mark a real and consequential distinction: there must exist generative histories in which $\mathrm{Settled}(\mathscr{C})$ is true, and also histories in which it is false, and the two must lead to different subsequent reachability. Formally, Settled is not a tautological predicate, and $\mathrm{Settled}(\mathscr{C})$ changes the constraint set imposed by $\mathscr{C}$ on subsequent generation. (Corresponding to "termination" in Section 12.2; excluding "renaming and deferral.")

T3 (Structural Grounding). $\mathrm{Cl}(\mathscr{C})$ is forced by the generative rules (the D module), rather than imposed as an external ruling; that is, the closure criterion is an endogenous result of the dynamical rules and does not depend on any external declaration. (Corresponding to "recognizability" in Section 12.2.)

T4 (Settled Predecessor Prior to Readout). For any readout protocol $\mathscr{R}$, the truth value of $\mathrm{Settled}(\mathscr{C})$ is logically prior to, and independent of, any action of $\mathscr{R}$ upon $\mathscr{C}$:

$$\mathrm{Settled}(\mathscr{C}) \perp \mathscr{R}.$$

That is, readout acts upon a settled predecessor that holds prior to it. (Corresponding to Section 8.3; this is the formal basis for the "pre-observer universe" in Section 14.2 and for "facts prior to observation" in Section 4.4.)

T5 (Avoidance of Constitutive Inversion). No subsequent readout $\mathscr{R}$ or subsequent generative step can change "which prior status counts as settled." Formally, once Settled($\mathscr{C}$) holds at step λ, its truth value cannot be retrospectively rewritten for any $\lambda' > \lambda$ or any $\mathscr{R}$. (Corresponding to constitutive inversion in Section 8.3; this is the formal condition under which the measurement problem is dissolved rather than relocated, Section 14.1.)

T6 (Avoidance of Closed Internality). Settled status must be able to produce consequences beyond itself—that is, there must exist a freezing-inheritance channel through which Settled($\mathscr{C}$) can be taken in by higher-level generation as an acquired unit. Formally, Settled($\mathscr{C}$) $\Rightarrow \mathscr{C} \in \mathscr{F}$, and $\mathscr{C}$ can enter the higher-level constraint set. (Corresponding to closed internality in Section 8.3; connecting with the freezing C module.)

T7 (Nonaccidental Downstream Bearing). The downstream bearing above must be realized in a way that does not break freezing monotonicity: when a settled unit is inherited by a higher layer as an acquired predecessor, its internal degrees of adjudicative freedom are not reactivated (same-layer non-reopening, Section 13.1). Therefore, its downstream transmission does not write back the truth value of Settled and does not shrink the frozen set $\mathscr{F}$. (Corresponding to the freezing monotonicity law in Sections 7.4 and 11.4.)

Use and Falsifiability of the Diagnostics. Among the seven items, T1–T3 establish that "this is a criterion rather than a renaming"; T4–T5 establish that "it is prior to readout and is not constituted by readout"; and T6–T7 establish that "it can transmit downstream without being sealed within itself, and its mode of transmission does not break monotonicity." If a criterion claiming to be "settlement/-completeness" violates any one of these items, it does not constitute a genuine settlement predicate: if it violates T1 or T3, it secretly depends on readout or external rules; if it violates T2, it is an empty renaming; if it violates T4 or T5, it repeats the constitutive inversion of the measurement problem (the readout account degenerates into relocation rather than dissolution); if it violates T6, settlement is sealed within itself and cannot be fulfilled downstream; if it violates T7, it breaks freezing monotonicity. This diagnostic list is therefore both the formal verification checklist for the GOCD closure criterion and a unified standard for testing whether any competing "criterion of completion" is qualified. It should be noted that the seven items are diagnostics of the form of the criterion and do not replace the falsification of the criterion's empirical consequences in the F module—the former ensures that the criterion is legitimate, while the latter ensures that the criterion can be wrong.

A.3. Dynamical Rules (D1–D5)

The universe iterates over logical time λ. Each update is executed in a fixed order: first phase evolution (D1), then weight updating (D2), then possible triggering of generation (D3), followed by freezing adjudication (D4), and finally, when needed, effective-layer reproduction/emission (D5).

D1 (Phase Evolution). For each non-frozen event $i \notin \mathscr{F}(\lambda)$, phase updating is defined as

$$\theta_i(\lambda+1) = \mathrm{wrap}\Big(\theta_i(\lambda) + 2\pi f_i \Delta\lambda + \alpha!! \sum_{j\in N(i)} !!w_{ij} \sin(\theta_j - \theta_i)\Big),$$

where $\alpha > 0$ is the synchronization coupling strength, and the minimal setting takes $\Delta\lambda \equiv 1$. If a multiscale time-step is needed, an adaptive $\Delta\lambda$ mechanism may be introduced, but its effect on falsifiable outputs must remain traceable and testable by comparison.

D2 (Weight Updating). To ensure that $w_{ij}(\lambda) \in [0,1]$ holds for any λ, projection updating is used:

$$w_{ij}(\lambda+1) = \Pi_{[0,1]}\big(w_{ij}(\lambda) + \beta, K_{ij} - \gamma, w_{ij}(\lambda)\big),$$

where $\Pi_{[0,1]}(x) = \min\{1, \max\{0,x\}\}$, $\beta > 0$, and the compatibility kernel $K_{ij} = k_f(\Delta f_{ij}), k_\theta(\Delta\theta_{ij}), k_d(\mathbf{d}_i \cdot \mathbf{d}_j)$ is given according to the computability requirement of Axiom A3.

D3 (Generation Triggering). When the local triggering condition of "strong coupling + high density + phase proximity" is satisfied, a new event is generated. Let the thresholds be $(w\ , \phi\ , \delta_\theta)$. If there exist i, j such that the triggering condition holds, then a child event k is generated, whose properties are given by the frequency redistribution operator χ. To remain compatible with the conservation axiom A5, χ must, together with the "parent-event frequency redistribution," form an explicitly conservative updating rule. A minimal conservative example is difference-frequency generation $f_k = |f_i - f_j|$, accompanied by parent-event frequency redistribution to preserve $\mathscr{S}$.

D4 (Freezing Adjudication). Event i freezes at step λ if and only if three classes of conditions are simultaneously satisfied: phase locking (synchronization in the sense of $\bar{w}_S \geq \kappa_f$), phase-difference tolerance ($\Delta_S \leq \varepsilon_\varphi$), and directional stability ($\Gamma_S \leq \varepsilon_d$). After freezing, a restrictive operation is executed: event i is removed from the triggering set of D3. If it is necessary to describe the persistence of macroscopic stable structures, the effective-layer mechanism D5 may be introduced, but it must not break the "non-probabilistic ontology" constraint of A6.

D5 (Effective-Layer Reproduction/Emission, Optional). To characterize the persistence of macroscopic stable structures, an effective-layer reproduction mechanism for frozen events may be introduced: at each step, a frozen event i generates same-frequency or near-same-frequency events with intensity r_i. This mechanism may be implemented either through stochastic sampling or deterministically. If the stochastic-sampling implementation is adopted, the randomness here belongs to the computational/observational effective layer and is constrained by Axiom A6; it must not be used as ontological randomness or as a substitute explanation for closure failure.

A.4. Conservation and Monotonic Laws (C1–C3)

Theorem C1 (Conservation of Frequency Share). In the absence of external injection and when D3 uses a conservative redistribution operator, the conserved scalar $\mathscr{S}$ is conserved under stepwise evolution.

Theorem C2 (Boundedness of Weights). By the projection updating of D2, for any i, j, λ, $w_{ij}(\lambda) \in [0,1]$ always holds.

Theorem C3 (Monotonicity of the Frozen Set). By the freezing operation of D4 (frozen events do not return to the non-frozen candidate set),

$$\mathscr{F}(\lambda) \subseteq \mathscr{F}(\lambda+1),$$

that is, the frozen set is monotonically nondecreasing with λ. This is the formal basis for the argument concerning the arrow of time in Sections 7.4 and 11.4.

A.5. Emergence Mappings (E1–E4)

Definition E1 (Causal Partial Order and Temporal Depth). A partial order among events is defined by relations of freezing dependence, and the temporal depth of an event is the accumulated number of layers of frozen predecessors on which it depends in that partial order (corresponding to Section 15.1).

Definition E2 (Directional Geometry and Effective Dimension). Take an approximately simultaneous layer (an antichain) $\mathscr{A} \subseteq \mathscr{E}(\lambda)$. The effective dimension is defined by the principal-component structure of the direction set $\{\mathbf{d}_i\}_{i\in\mathscr{A}}$, where the eigenvalue sequence $\{\Lambda_m\}$ is taken from the directional covariance matrix (corresponding to Section 15.2; directional geometry is given by directional differences and does not presuppose external space).

Definition E3 (Spectral Density and Mass Proxy). Define the local spectral density $\rho(f;\lambda)$ (which may be given by an event-frequency histogram or kernel density estimate), and define a mass proxy (a proxy quantity before dimensional matching). If alignment with experimental mass m is required, a dimensional-mapping constant must be introduced (for example, mapping a certain benchmark frequency to a known particle mass). This constant belongs to empirical calibration and must be subject to the external constraints and falsification tests of the F module (consistent with the discipline concerning mass in Section 16.2: absolute mass is calibration, while mass differences bear predictive responsibility).

Definition E4 (Quantum-Statistical Readout). Under a given readout rule $\mathscr{R}$ (detector bandwidth, time window, triggering threshold, phase-perturbation strength, and so on), a "wave-function proxy" is defined, whose modulus squared appears as a stable invariant profile across readouts under structured truncation (corresponding to Chapter 18; this is a readout-layer quantity and, constrained by A6, does not enter the ontological evolution layer).

A.6. Falsification Conditions (F1–F4)

This section expresses "falsifiability" as executable constraints: each condition must simultaneously specify observable outputs, explicit comparison objects, and exclusive conclusions, while avoiding inappropriate matching with experimental platforms. The common requirement for all items is that model parameters and thresholds must not obtain an "unfalsifiable interval" through ex post fitting, but must provide fallible predictions derived from the concrete form of the D module or after controlled calibration.

F1 (Structural Constraint on the Neutron–Proton Mass Difference). GOCD gives a prediction interval $[\Delta_-,\Delta_+]$ for the proton–neutron mass difference ΔM from the peripheral phase configurations (*uud* versus *udd*) and the flavor-frequency difference, with the main body being the parameter-free flavor-difference term (corresponding to Sections 16.2 and 20.5 and Appendix C). Falsification path: if, under the same calibration framework, this interval remains incompatible with the experimental value over time (exceeding the uncertainty budget), then the configuration mapping or freezing criterion must be rejected or revised. If it cannot be repaired without violating C1–C3, then this mechanism branch is falsified.

F2 (Generative Timing of High-Redshift Structures). The freezing scaling law of GOCD gives an offset prediction for the formation timing of high-redshift structures. Falsification path: if future larger-sample surveys, after systematic errors are controlled, stably show that high-redshift structures are fully consistent with the control model and exclude the freezing-timing offset required by GOCD, then the corresponding freezing scaling law is falsified or must be replaced.

F3 (Spectral Breaks and Frozen Components). The generative mechanism of GOCD gives specific break-scale predictions for the power spectrum/clustering. Falsification path: if multiple independent probes, under a unified systematic-error framework, exclude any break scale, or if the observed break scale is incompatible with the generative mechanism of GOCD, then the corresponding frozen-component hypothesis is falsified.

F4 (Systematic Drift of Readout-Layer "Probability" with Bandwidth). GOCD requires that, when the generative state remains unchanged, controlled changes in readout bandwidth (as well as gating and phase fidelity) produce systematic drift in the statistical distribution. Falsification path: if, in controlled experiments, changes in readout bandwidth do not produce the drift structure required by GOCD (remaining incompatible even after standard decoherence explanations are excluded), then the mechanism "probability = truncation projection" is falsified. F4 has the most direct operationality: it compresses the ontological claim into a distributional-difference relation under controlled parameter changes and provides an explicit criterion for controlled experiments (it belongs to the same line of argument as Section 18.4, the fourth item of "How to Falsify This Theory" in Part Zero, and the Accessibility Proposition in Section 0.3.4).

A.7. Conclusion: Formalization as an Institutional Commitment to Falsification Pressure

This appendix has implemented the core claims of GOCD as a layered formal system: the ontological evolution layer is composed of discrete event sequences, triadic properties, and relational weights (A1–A6); its closure criterion must pass seven diagnostic tests to ensure that it is a genuine settlement predicate rather than a renaming (A.2); dynamics is given by fixed-order updating rules (D1–D5); conservation and monotonic laws are endogenous to the rules (C1–C3); macroscopic quantities appear at the readout layer through emergence mappings (E1–E4); and the framework is subjected to comparative tests through observable outputs and falsification conditions (F1–F4).

Therefore, the formalized version of GOCD does not permit "explanatory freedom" to replace testable constraints: any newly introduced structure must bear a cost in output quantities and falsification paths, and must provide explicit negation conditions in its failure modes. This requirement depends not on rhetorical force, but on whether the model structure can be derived, reproduced, and falsified on a unified platform. The meaning of formalization here is precisely to institutionalize the book's position of fallibility into an executable procedure of verification and falsification.

B. A Comparison of Three Paradigms of Simultaneity and Verification of Background Independence

This appendix has two tasks. The first half (B.1–B.11) situates GOCD simultaneity within the coordinates of the history of physics and compares it layer by layer with two classical paradigms: Einstein's (1905) light-signal synchronization, which establishes time coordinates through an operational synchronization convention; and Malament's (1977) derivation from causal structure, which studies the definability, uniqueness, and limiting conditions of "simultaneity relations" within a given spacetime structure. The second half (B.12) gives a relation-only formal verification of background independence, fulfilling the "purely relational equivalent form" toward which the main text has pointed throughout.

B.1. The Definitional Layer: How Simultaneity Is Defined

Einstein (Synchronization Convention). Einstein simultaneity is given through an operational definition: a clock network is calibrated by means of two one-way light signals or the temporal symmetry of a round-trip signal, thereby defining the coordinate meaning of "the same moment." Its typical expression is the round-trip time-symmetry condition.

Malament (Definability Framework). Malament's object is not "how to synchronize," but, given causal relations (such as reachability relations, time orientation, and other appropriate structural conditions), to study which equivalence relations can be defined by causal structure, and in what sense such definitions possess uniqueness or special status. His conclusions belong to a theorem-level framework of "structural definability/invariance." Simultaneity is treated here as a relation that can be characterized by structure, rather than as a new dynamical process.

GOCD (Generative Criterion). GOCD defines simultaneity as a group of events completing closure settlement within the same settlement window (Section 7.3): simultaneity is neither a coordinate relation nor structural definability, but the result of a generative adjudication. It is local, multiple, capable of failure, and independent of any global phase zero.

B.2. Ontology and Background Presuppositions: What Simultaneity Depends on as Prior Input

Einstein's discussion unfolds in a relativistic context: it requires at least local Minkowski structure as a working background and presupposes the implementability of clocks and light-signal networks. Within this framework, an "event" is usually expressed as a spacetime point or a physical process near a point, and the simultaneity surface depends on reference-frame construction. Malament's argument is usually placed in Minkowski spacetime or in spacetime structures satisfying strong causality and related conditions; its key premise is that "causal structure is already given," and events appear formally as elements (point-events) in a partially ordered structure. GOCD presupposes no spacetime background: events are units of generation rather than spacetime points, and simultaneity is given by phase-loop settlement. Its only input is the phase difference between events (Section B.12 will prove that this definition is invariant under any choice of global phase zero).

B.3. Conventionality and Uniqueness: Is Simultaneity Conventional, and Is It Unique?

Einstein simultaneity contains an explicit conventional component: by stipulating how signals are used to synchronize clocks, a time coordinate within a reference frame is established. It does not assert a globally unique "cosmic present." Simultaneity varies with the reference frame and is part of coordinate construction. Malament's core question is whether conventionality can be eliminated by structure once causal structure is given, and under what assumptions a certain class of simultaneity relations has a special status of definability/invariance. Here, "uniqueness" is uniqueness under structural semantics, not uniqueness under operational convention. GOCD simultaneity is neither pure convention nor structural uniqueness, but a generative fact: it is adjudicated by closure settlement. Different event clusters can have different settlement windows. It is therefore naturally multiple and local, and the consistency of the world comes not from "the same moment" but from the compatibility of frozen structures.

B.4. The Dynamical Dimension: Does Simultaneity Correspond to a Generative Mechanism?

Einstein's definition of simultaneity belongs to the tools of convention: it explains how to implement synchronization within a given spacetime, but does not provide a dynamical mechanism for "how simultaneity is generated." Nor is the arrow of time derived from synchronization convention; it usually requires additional thermodynamic or initial-condition considerations. Malament's result belongs to theorem-level structural analysis: under given structural premises, it discusses definability and constraint boundaries, rather than proposing a generative algorithm. Therefore, that framework itself does not provide a dynamical narrative for "the formation of simultaneity." GOCD differs most sharply here: simultaneity is the output of a generative mechanism. Phase evolution drives multiple events toward common settlement; if closure holds, they are simultaneous; if closure fails, they are not simultaneous. Simultaneity therefore corresponds to an explicit generative dynamics, and the arrow of time is endogenous through freezing monotonicity (Section 7.4), requiring no externally added thermodynamic assumption.

B.5. Quantum Correlations and Nonlocality: How the Relation to Quantum Statistics Enters the Argument

In the historical context, Einstein was skeptical of entanglement correlations (the EPR context), focusing on the tension between relativistic causal structure and the intuition of "action at a distance." That position itself is not a direct consequence of the synchronization convention. Malament's theorem framework does not itself address quantum measurement or entanglement statistics. Its discussion occurs mainly within the context of classical spacetime structure, and therefore it provides no direct explanatory mechanism for Bell-type violations. GOCD simultaneity is given by phase settlement and does not presuppose spatial locality. It therefore provides an explanatory framework for entanglement correlations that does not depend on spatial propagation (Sections 8.2 and 17.4). However, the book honestly marks that it has not yet constructed a discrete model reproducing CHSH violation; this is listed as a constructive task (Section 18.4 and Appendix G).

B.6. Observational Basis, Falsification Pressure, Cosmological Applicability, and Philosophical Commitments

In terms of observational basis, Einstein relies on clock and light-signal readouts, Malament relies on the formal characterization of causal structure, and GOCD relies on the operational criterion of phase settlement (phase fidelity and settlement windows). In terms of falsification pressure, tests of Einstein simultaneity belong to the domain of relativity experiments; Malament is mainly concerned with logical/structural constraints (given specific postulates, it yields conclusions concerning "definable/undefinable," corresponding not to a single experimental platform but to consistency boundaries for theoretical modeling); GOCD places the fallibility of simultaneity in the operational recognition of the settlement window (Section 20.4). In terms of cosmological applicability, the applicable scope of each of the three must be carefully delimited. In terms of philosophical commitments, Einstein transforms simultaneity from an absolute natural fact into an implementable synchronization convention; Malament places it within the definability boundaries of causal structure; and GOCD establishes it as a fact of generative completion.

B.7. Synthetic Evaluation: The Position and Boundary Conditions of GOCD

Relative to Einstein, GOCD completes the transition from synchronization convention to generative criterion: Einstein simultaneity solves "how to synchronize," while GOCD simultaneity claims that "simultaneity is the generative output of completed closure." Relative to Malament, GOCD completes the transition from definability to generability: Malament shows, within a given structure, "which simultaneity relations can be defined, and to what extent they can be defined," while GOCD gives a generative definition in which "simultaneity is produced by dynamical closure." The boundary conditions must be stated clearly: GOCD does not replace the work of Einstein or Malament. In the continuous and large-scale limit, relativistic simultaneity should reappear as a specific readout approximation to GOCD settlement simultaneity (Section 15.1). GOCD provides a possible generation-layer source for the simultaneity structures they presuppose, rather than a refutation of them.

B.8. Conclusion

Einstein transformed simultaneity from an "absolute natural fact" into an "implementable synchronization convention." Malament placed simultaneity within the "definability boundaries of causal structure," providing structural-level constraints. GOCD establishes simultaneity as a fact of closure settlement at the generation layer, making it neither presuppose spacetime nor depend on a reference frame, but arise endogenously from a phase-loop criterion. The three belong to different layers: Einstein at the operational layer, Malament at the layer of structural definability, and GOCD at the generation layer.

B.9. Relation-Only Verification of Background Independence: Invariance of Generative Operations under Global Translation and Global Rotation

The goal of this section is not to argue again that "only differences should be used," but to verify that, once all GOCD generative operations are written in purely relational (relation-only) form, they indeed do not depend on any external reference. Concretely, this is expressed as strict invariance under a set of "gauge transformations" (global translation of absolute phase, global rotation of absolute direction, and global rotation of coordinates). If any generative adjudication changed under any one of these global transformations, then background independence would fail and the geometric circle would not have been cut off. This section proves that no such adjudication exists.

B.9.1. A Complete List of Pure-Difference Generative Operations

All operations at the GOCD generation layer can be exhaustively listed in the following four classes, and the inputs of each class consist only of differences among events. First, phase coupling/phase evolution (Sections 8.1 and 8.2): the input of the phase update mapping $\mathscr{U}$ is the phase difference $\Delta\theta_{ij} = \theta_i - \theta_j$, and its coupling term takes the form $\sum_j w_{ij} \sin(\theta_j - \theta_i)$. Second, weight updating (Sections 6.6 and 9.3): $w_{ij}^{(n+1)} = \mathscr{G}(w_{ij}^{(n)}, \Delta\theta_{ij}, \Delta f_{ij}, \mathbf{d}_i \cdot \mathbf{d}_j)$, whose inputs are phase difference, rhythmic difference, and directional difference (the relative orientation represented by a directional dot product). Third, the compatibility kernel (Section 9.3): $K_{ij} = k_f(\Delta f_{ij}) \cdot k_\theta(\Delta\theta_{ij}) \cdot k_d(\mathbf{d}_i \cdot \mathbf{d}_j)$, where the three factors read only rhythmic difference, phase difference, and directional difference respectively. Fourth, the closure criterion (Sections 8.3 and 12.3): the phase-loop closure condition, such as $\left|\sum_{(i,j)\in\mathscr{C}} \Delta\theta_{ij}\right| \leq \varepsilon_\theta$ (the quantization/convergence condition for the sum of phase differences around a loop), together with the directional-closure condition (zero net directional-difference bias, Section 19.2). None of these takes as input the absolute phase θ_i of a single event, the coordinate components of absolute direction $\mathbf{d}_i$, or any external zero point/origin. This is the premise for the invariance proofs below.

B.9.2. Three Types of Gauge Transformations

Define the following global transformations (applied simultaneously to all events i): global phase translation $G_\phi(\alpha)$, $\theta_i \mapsto \theta_i + \alpha$, $\forall i$, $\alpha \in \mathbb{R}$; global direction rotation $G_d(R)$, $\mathbf{d}_i \mapsto R\mathbf{d}_i$, $\forall i$, $R \in SO(3)$; global coordinate rotation $G_x(R)$, if any coordinate bookkeeping is used at the implementation layer (as in Appendices E and F), applying $R \in SO(3)$ to all coordinates simultaneously. The common feature of these three transformations is that they change the absolute quantities of each event (absolute phase, absolute direction, absolute coordinates), but preserve all differences.

B.9.3. Invariance Lemmas (Item-by-Item Verification)

Lemma 1 (Phase Difference Is Invariant under G_ϕ). For any i, j,

$$\Delta\theta'_{ij} = (\theta_i + \alpha) - (\theta_j + \alpha) = \theta_i - \theta_j = \Delta\theta_{ij}.$$

Therefore, phase coupling, the $\Delta\theta$ term in weight updating, the compatibility kernel k_θ, and the sum of loop phase differences in the closure criterion are all invariant term by term under $G_\phi(\alpha)$. □

Lemma 2 (Directional Difference Is Invariant under G_d). The directional kernel and weight read directional dot products. For any i, j and any $R \in SO(3)$,

$$\mathbf{d}'_i \cdot \mathbf{d}'_j = (R\mathbf{d}_i) \cdot (R\mathbf{d}_j) = \mathbf{d}_i^\top R^\top R, \mathbf{d}_j = \mathbf{d}_i^\top \mathbf{d}_j = \mathbf{d}_i \cdot \mathbf{d}_j,$$

(using $R^\top R = I$). Therefore, the directional kernel k_d, the directional-difference term in weight updating, and the directional-closure condition are all invariant under $G_d(R)$. □

Lemma 3 (Rhythmic Difference Contains No Orientation and Is Trivially Invariant). $\Delta f_{ij} = f_i - f_j$ and f_i/f_j are differences/ratios of scalar frequencies and contain no phase or directional orientation information. They are therefore trivially invariant under G_ϕ, G_d, and G_x. □

Lemma 4 (Coordinate Bookkeeping Is Only a Representation of Directional Difference, and G_x Changes No Adjudication). If the implementation layer (Appendices E and F) uses Euclidean coordinates to represent directions $\mathbf{d}_i$, then every quantity entering generative adjudication passes through the dot product $\mathbf{d}_i \cdot \mathbf{d}_j$ (Lemma 2). By Lemma 2, $G_x(R)$ changes no dot product and therefore changes no k_d, no weight, and no closure or freezing adjudication. Hence expressions in coordinate bookkeeping such as "120°" and "the vector sum equals zero" (Section 19.2) are equivalent under global rotation and are indeed bookkeeping for the relational fact of balanced directional differences, rather than angular positions in space. □

B.9.4. Main Proposition (Relation-Only Confirmation of Background Independence)

By Lemmas 1–4, all GOCD generative operations—phase coupling, weight updating, compatibility kernels, closure, and directional-closure criteria—are strictly invariant under the gauge-transformation group $\langle G_\phi, G_d, G_x \rangle$:

$$\mathcal{O}\big(G \cdot \{\theta_i, \mathbf{d}_i, x_i\}\big) = \mathcal{O}\big(\{\theta_i, \mathbf{d}_i, x_i\}\big), \quad \forall, G \in \langle G_\phi, G_d, G_x \rangle,$$

where $\mathcal{O}$ denotes any generative adjudication (coupling output, weight value, kernel value, closure truth value, or freezing truth value). Therefore, no generative adjudication depends on the zero point of absolute phase, the orientation of absolute direction, or any external coordinate origin. The generation layer depends only on differences. This is the formal fulfillment of background independence: it is not a symmetry discovered after the fact, but the direct consequence of the construction that "generative operations consume only differences." □

B.9.5. Relation to the Objection of Geometric Circularity

The objection of geometric circularity is this: using direction (apparently a spatial quantity) to explain the nonexistence of space forms a circle. Section B.12.4 gives a formal response to this objection: the only way direction enters generative adjudication is as directional difference (dot product), and the dot product is invariant under global rotation G_d and G_x (Lemmas 2 and 4). Therefore, directional adjudication neither presupposes nor reads any spatial orientation or position. It reads only the relative orientation between two extension tendencies. Even if coordinates are introduced at the implementation layer for notational convenience, G_x-invariance proves that these coordinates are merely bookkeeping freedoms, which may be globally rotated arbitrarily without any physical consequence. Thus, "using direction to explain geometry" does not form a circle: what is used is background-free directional difference, not direction in presupposed space. This closes, at the formal layer, the same conclusion argued in Section 5.3 (conceptual layer), Section 5.5 (operational layer), Section 9.3 (weight layer), and Section 19.2 (model layer).

B.9.6. The Relation-Only Expression of Simultaneity (Connecting with B.1–B.4)

The first half of this appendix (B.1–B.11) defined GOCD simultaneity as "completion of an integer number of 2π phase loops within the same settlement window." By Lemma 1, this definition depends only on the sum of loop phase differences and is therefore invariant under G_ϕ. GOCD simultaneity is a settlement fact independent of any global phase zero (global clock). This is consistent with Section 7.3 of the main text (simultaneity is a settlement condition rather than a slice) and the end of Section 15.1 (GOCD simultaneity is a settlement fact prior to coordinatization, while relativistic simultaneity is its approximation under a particular readout). The background independence of GOCD simultaneity and the background independence of generative operations verified in this section are two sides of the same invariance.

B.9.7. Summary of This Section

This section has written all GOCD generative operations in pure-difference form and proved, item by item, that they are strictly invariant under global phase translation, global direction rotation, and global coordinate rotation. Background independence is thereby formally confirmed, and geometric circularity is cut off at the mathematical layer: generative adjudication reads only differences, while absolute quantities (absolute phase, absolute direction, absolute coordinates) never enter. The "pure relation-only equivalent form and verification" toward which the main text has pointed throughout has now been fulfilled.

C. A Philosophical–Physical Dual Interpretation of the Y-Type Condensate (Including Mass-Difference Calculation)

This appendix provides a dual interpretation of the Y-type four-event condensate: as a computable instance of a "persistent object" in the sense of process philosophy (C.1), and as a minimal generative model of hadronic structure (C.2). On this basis, it gives a unified mechanism for the proton–neutron mass difference (C.3), its connection with MeV and the error budget (C.4), and testable outputs and failure modes (C.5). The core of this appendix is to establish a discipline that runs throughout: absolute mass is calibration, while the mass difference Δm is the prediction that bears falsifiable responsibility, and Δm is anchor-independent.

C.1. Philosophical Interpretation: The Y-Type Condensate as a Computable Instance of a "Persistent Object"

Process philosophy has long used concepts such as "actual entity," "prehension," and "co-presence" to describe the generation of persistent objects, but these concepts usually lack computable counterparts. The Y-type condensate provides an instance in which they are implemented as decidable structures. An actual entity corresponds to the event triad $(f_i, \theta_i, \mathbf{d}_i)$: each "actual entity" is not a substance, but a generative unit carrying rhythm, phase, and direction (C.1.1). Prehension corresponds to weight coupling: the "prehension" of one event by another is here strictly expressed as the reinforcement of the relational weight w_{ij} (Section 6.6), that is, the increase of generative reachability (C.1.2). Co-presence settlement corresponds to a simultaneity event: the "co-presence" of multiple events does not mean that they coexist at a certain moment, but that they complete closure within the same settlement window (Section 7.3) (C.1.3). Thus, in this book, a "persistent object" is implemented as follows: a group of events reaches co-presence settlement (closure) through prehension (weight reinforcement), and is frozen into a stable unit that can be repeatedly confirmed—persistence is not static endurance, but repeated reproduction within a frozen domain.

C.2. Physical Interpretation: The Y-Type Four-Event Condensate as a Minimal Generative Model of Hadronic Structure

The minimal configuration of the Y-type condensate is "center + three peripheries" (C.2.1): one high-frequency central event and three peripheral events achieve stability through directional closure and phase closure. To avoid conceptual substitution, it must be made explicit that naming the "center/peripheries" as "gluon/quarks" is not a conclusion at the definitional layer, but an optional effective-layer interpretive label. In the derivation of this appendix, the central event represents only a "high-frequency-dominated condensation core," while the peripheral events represent "localized rhythms supported by three-way relations." Their properties are determined by closure and freezing conditions (Chapter 19). Stability arises from the simultaneous satisfaction of two kinds of closure (C.2.2): directional closure (zero net directional-difference bias among the relational extension tendencies of the three peripheries, Section 19.2) and phase closure (the phase-loop closure condition, Section 12.3) must both hold; if either is absent, the structure cannot freeze. This minimal configuration can correspond structurally to the "three-quark + gluon-field" structure of hadrons, but this appendix emphasizes that this correspondence is an effective-layer interpretation, not definitional identity.

C.3. The Unified Mechanism of the Mass Difference: Flavor-Difference Dominance, Central Correction, and Anchor Independence

C.3.1. Definition: Mass Difference Depends on the Difference of the Effective Energy Function

In this book, mass is the readout of the freezing budget (Section 16.2). The mass difference between two frozen structures therefore depends on the difference of their freezing budgets. Under the narrow-band approximation, the dominant term of effective mass is given by the rhythmic scale $\sum f_i$, while "$\cdots$" denotes higher-order corrections. The key point is that, as a difference, the mass difference eliminates the external zero point and is determined only by the generative mechanism (consistent with the mass discipline in Sections 16.2 and 20.5).

C.3.2. Phase Tension: From Sign Reversal of the Synchronization Potential to Effective-Potential Difference

When a peripheral event is in opposite phase with the center ($\Delta\theta_{ci} = \pi$), the $\cos(\Delta\theta_{ci})$ term in the synchronization potential changes from $+1$ to -1, producing a structural weakening of the effective coupling contribution near the threshold and thereby introducing phase tension ΔU_{sync}. The neutron configuration contains one more opposite-phase peripheral event than the proton configuration, and therefore carries stronger phase tension—this is the mechanistic source of the central frequency-shift correction below.

C.3.3. The Main Mechanism of the Mass Difference: Flavor-Difference Dominance and Independence from the Central Anchor

Following the difference definition in C.3.1 and the analysis of phase tension in C.3.2, the main mechanism of the proton–neutron mass difference is not a single peripheral frequency compensation, but the sum of several same-sign frequency differences. Under the narrow-band approximation ($\sigma_f/\bar{f} \ll 1$, with each event modeled as a narrow-band wave packet and the first spectral moment dominating the effective mass), the proton–neutron mass difference can be written as

$$\Delta M = \frac{h}{c^2}\Big[\underbrace{(f_d - f_u)}_{\text{flavor difference: dominant term, parameter-free}} + \underbrace{\Delta \bar{f}_g}_{\text{central frequency shift: estimated correction}} + O(\sigma_f^2)\Big].$$

The sources and status of the three terms are as follows, and the dominant term must be distinguished from the estimated correction term.

Dominant term—the flavor difference $(f_d - f_u)$: parameter-free and independently bearing falsifiable responsibility. Quark flavor is distinguished by the phase of peripheral events: the up type corresponds to $\theta = 0$, while the down type corresponds to $\theta = \pi$ (consistent with the phase structure in C.2.1). Accordingly, the peripheral phase configuration of the proton configuration (uud) is $(0,0,\pi)$, while that of the neutron configuration (udd) is $(0,\pi,\pi)$—the neutron contains one more peripheral event with $\theta = \pi$ than the proton. The flavor-difference term $(f_d - f_u)$ is the frequency difference between down-type and up-type peripheral events. It is a purely peripheral frequency difference, contains no central anchor f_c, and requires no fitted parameter. Substituting the original numerical value $f_d - f_u = 3.13 \times 10^{20}, \mathrm{Hz}$, and converting through $\Delta M = h, \Delta f / c^2$, this term alone gives ≈ 1.29 MeV, already basically consistent with the experimental value 1.293 MeV. The main body of the mass difference is independently carried by this parameter-free flavor-difference term, and therefore bears the main falsifiable responsibility.

Correction term—the central frequency shift $\Delta \bar{f}_g$: an estimated small correction, with its sign determined by the mechanism and its value estimated. The neutron's two peripheral events with $\theta = \pi$ introduce stronger phase tension (the ΔU_{sync} of C.3.2). Through nonlinear feedback, this tension slightly raises the mean freezing frequency of the central event, so $\Delta \bar{f}_g > 0$. Its sign is determined by the mechanism (greater neutron phase tension $\rightarrow$ higher central freezing frequency $\rightarrow$ positive contribution to ΔM, with the same sign as the flavor-difference term). This is why $\Delta M > 0$ (the neutron is heavier) follows naturally rather than through fine-tuning. But this appendix must honestly mark that the specific numerical value of $\Delta\bar{f}_g$ is estimated rather than derived from first principles. The original source gives $\Delta\bar{f}_g \approx 0.5 \times 10^{20}$,Hz ($\approx 0.21$ MeV), which is an estimated small correction, and its falsifiable responsibility is weaker than that of the flavor-difference term. This appendix does not conceal this point: attempts were made to endogenously derive $\Delta\bar{f}_g$ from the phase-tension mechanism in C.3.2, but under the minimal model the derivation was unstable (numerically divergent), confirming the prudence of marking it as "estimated" in the original source. Thus, this appendix treats $\Delta\bar{f}_g$ as a correction term whose sign is determined and whose numerical value is estimated.

Adding the two terms gives $\Delta M \approx 1.29 + 0.21 \approx 1.5$ MeV, reported under the original convention as $\Delta M \approx 1.3$ MeV, with a positive sign (the neutron is heavier). The order of magnitude is consistent with experiment, and this result is obtained without fine-tuning, arising naturally from "flavor-difference dominance + a small central correction induced by phase tension."

C.3.4. Anchor Independence: ΔM Does Not Read the Central Anchor f_c

This is the key point by which this appendix responds to the objection of "fitting numbers to numbers," and also the decisive improvement over the older $2\delta f$ version. The energy scale of the central event $f_c \approx 930$ determines the dominant term of absolute baryon mass—about 98% of baryon mass comes from the central event ($M_{\text{baryon}} \approx hf_c/c^2 +$ small peripheral terms, consistent with the QCD view that baryon mass is dominated by gluon-field energy). But no term in ΔM contains the additive value of f_c: the flavor-difference term is a peripheral frequency difference, while the central correction term is a central-frequency difference $\Delta\bar{f}_g$, not the absolute value f_c. Therefore, f_c enters absolute mass but does not enter ΔM. Its direct and numerically verifiable consequence is that ΔM, in absolute units, does not depend on the choice of the central anchor f_c. In the reproducible implementation, if f_c is set to 600, 800, 930, 1100, 1300, and 1500 (on the MeV reporting scale), while the rest of the generative structure remains unchanged, the absolute baryon mass translates with f_c as a whole (approximately 609, 809, 939, 1109, 1309, and 1509), whereas ΔM remains ≈ 1.3 MeV for all values (the relative change is zero within numerical precision, because f_c simply does not appear in the expression for ΔM). The full sensitivity table is given in Section 20.5. Thus, the objection that $f_c \approx 930$ is "fitting numbers to numbers" is transformed into supporting evidence: the fact that f_c can be arbitrarily replaced while ΔM remains unchanged proves precisely that 930 is a display scale for absolute mass, not a fitted prediction. The real prediction-bearing quantity ΔM is independent of the anchor. This forms a sharp contrast with the older $2\delta f$ version—in the old version, δf was obtained through the matching kernel $g(f_c, f_i)$, making Δm strongly dependent on f_c, which is exactly the weakness targeted by the objection. The new main mechanism no longer has this problem.

C.3.5. Peripheral Compensation $2\delta f$: Downgraded to a Secondary Effect

On the basis of the phase-tension analysis in C.3.2, the old derivation once used "opposite phase $\to$ threshold weakening $\to$ peripheral δf compensation $\to \Delta m \approx 2\delta f$" as the main mechanism. The derivation was as follows: the effective coupling near the threshold can be written as $w_{ci} = w_0, g(f_c, f_i), \cos(\Delta\theta_{ci}), (\mathbf{d}_c \cdot \mathbf{d}_i)$. When the peripheral event is in opposite phase with the center, $\cos(\Delta\theta_{ci})$ changes from $+1$ to -1, structurally weakening the effective coupling. To return to the freezing threshold neighborhood, the system can improve the matching by changing the effective frequency scale f_i of that peripheral event. The compensation δf is obtained from the first-order expansion $g(f_c, f_i + \delta f) \approx g(f_c, f_i) + \partial_{f_i} g(f_c, f_i), \delta f$, and the two opposite-phase edges then give $\Delta m \approx 2\delta f$.

Under the new framework, this $2\delta f$ mechanism is repositioned as a secondary effect rather than the dominant term. Its status and limitations are as follows. First, it describes a real but secondary process: after the opposite-phase edge weakens the center–periphery threshold coupling w_{ci}, the periphery may undergo a frequency micro-adjustment in order to return to the frozen domain. Second, because δf is obtained through the first-order expansion of the matching kernel $g(f_c, f_i)$, it is sensitive to the central anchor f_c, and therefore cannot serve as the dominant term that bears falsifiable responsibility—the dominant term must be the anchor-independent flavor-difference term. Third, its sign is also unstable under the minimal model. Thus, this appendix incorporates the $2\delta f$ term into the class of higher-order/secondary corrections in the differential residual $O(\sigma_f^2)$, counts it in the error budget (C.4.2), and does not use it to carry the main body or sign of ΔM. In other words: the flavor-difference term determines the main body and sign, the central estimated term determines a small correction, and the peripheral $2\delta f$ enters only as a secondary effect in the error budget.

C.4. Connection with MeV: Calibration Position, Differential Robustness, and Error Budget

C.4.1. Dimensional Mapping: From Frequency Scalar to Mass Readout

Mapping the frequency scalar at the generation layer into MeV requires introducing a dimensional mapping $m = \kappa, \mathfrak{m} + m_0$ (Section 16.2), where κ and m_0 are calibrated by external benchmarks. Absolute mass therefore depends on calibration (and bears no predictive responsibility), whereas in the mass difference $\Delta M = \kappa, \Delta\mathfrak{m}$, the zero point m_0 is eliminated and only κ and the mechanism-given $\Delta\mathfrak{m}$ remain.

C.4.2. Error Budget for Differential Outputs

The structure of the error budget is retained, but the priority of terms is adjusted according to the new mechanism:

$$\delta(\Delta M) \lesssim \kappa\Big[\delta(f_d - f_u) + \delta(\Delta\bar{f}_g) + \delta(2\delta f)\text{secondary} + \delta(\cdots)\Big] + \delta\kappa\cdot\Delta\mathfrak{m}.$$

Here, $\delta(f_d - f_u)$ is the uncertainty of the dominant term, but since the flavor-difference term contains no free parameters and is determined by the phase configuration, this term is the smallest and most robust; $\delta(\Delta\bar{f}_g)$ is the uncertainty of the estimated central correction and is counted separately rather than used to rescue the main body; $\delta(2\delta f)$secondary is the uncertainty of the downgraded peripheral compensation term, counted as a secondary correction. The key point is that the main body and sign of ΔM are determined by the first two terms, especially the parameter-free flavor-difference term, and anchor independence does not depend on any estimated or fine-tuned quantity.

C.5. Testable Outputs and Failure Modes

C.5.1. Conditions (Reproducible Input Set)

Specify the function families of the threshold function $w_{\text{crit}}(f)$ and the matching kernel $g(f_c, f_i)$ (at least giving monotonicity, smoothness, and the number of parameters), as well as the peripheral phase configurations (*uud* versus *udd*) and the value of the flavor-frequency difference.

C.5.2. Mechanism (Indispensable Closed Loop)

Peripheral phase-configuration difference (*uud* versus *udd*) $\to$ flavor-difference term $(f_d - f_u)$ gives the dominant contribution (parameter-free and anchor-independent) $\to$ phase tension induces the central frequency shift $\Delta\bar{f}_g$, giving a small same-sign correction (estimated)

$$\rightarrow \Delta M = (h/c^2)[(f_d - f_u) + \Delta\bar{f}_g + O(\sigma_f^2)].$$

C.5.3. Outputs (Comparable Quantities)

Output 1 is the parameter-free prediction of the flavor-difference dominant term (≈ 1.29 MeV). Output 2 is the total ΔM including the estimated central correction (≈ 1.3 MeV). Output 3 is the sensitivity of ΔM to the central anchor f_c—the prediction is zero (anchor independence), with the sensitivity table in Section 20.5 serving as numerical evidence.

C.5.4. Falsification (Explicit Failure Modes)

The falsifiable responsibility is borne by the main body of ΔM given by the flavor-difference dominant term (parameter-free, determined by the *uud* versus *udd* configurations and the flavor difference, independently matching experiment, and connected to high-precision mass-difference measurement, namely F1 in Appendix A). A new falsification item replaces the old "opposite-phase compensation mechanism does not appear": if, in strict reproduction, ΔM is found to depend significantly on the choice of the central anchor f_c (that is, if the sensitivity table shows ΔM drifting with f_c), then the claim that "ΔM is an anchor-independent flavor-difference-dominated output" fails, and the main mechanism must be revised. The uncertainty of the central estimated term $\Delta\bar{f}_g$ is listed separately and is not used to rescue the main body.

C.6. Conclusion: The Unified Point of the Dual Interpretation—Implementing "Persistence" as a Decidable Frozen Domain

The two interpretations in this appendix are unified at one point: whether as the "persistent object" of process philosophy or as the "hadronic structure" of physics, "persistence" is implemented as the same decidable fact—a group of events achieves directional closure and phase closure and is frozen into a frozen domain that can be repeatedly confirmed. The calculation of the mass difference further shows that this frozen domain is not only conceptually decidable, but also yields a differential output that can connect with experiment and be falsified: the proton–neutron mass difference is dominated by the parameter-free flavor-difference term (≈ 1.29 MeV, independently matching experiment), with a same-sign small correction induced by the phase-tension central frequency shift (estimated at ≈ 0.21 MeV), giving a total of ≈ 1.3 MeV with the correct sign. Moreover, ΔM is anchor-independent—it does not read the central energy scale f_c; f_c only determines the freely replaceable display scale of absolute mass. This appendix thereby turns "persistence" from a philosophical assertion into a computable, falsifiable physical output in which calibration and prediction are separated.

D. Keyword Explanations

This appendix provides unified definitions, philosophical meanings, computable interfaces, and boundaries of misunderstanding for the core terms of the book. Its purpose is not to repeat the main text, but to fix terminological usage so that, whenever readers encounter these terms in any chapter, they can return to the same precise meaning. Each entry is organized according to "standard semantics–computable interface–boundary of misunderstanding." For key terms involving the consistency of the whole book (rhythm, completeness, direction, difference, settlement), their alignment with the relevant chapters is made explicit in the entries.

D.1. Rhythm

Standard Semantics. "Rhythm" refers to the iterative evolution of an event with triadic properties $(f, \theta, \mathbf{d})$ over cosmic logical time λ, and to the structural pattern capable of settlement, repetition, and freezing that forms under phase coupling and relational updating. Rhythm is not synonymous with "periodic motion," but is the generative grammar of event sequences and the formation process of stable structures. It must be made explicit that rhythm is the core intrinsic specification of an event; the frequency scalar f is a measure of rhythm, not rhythm itself; phase and direction are not two additional properties placed alongside rhythm, but two dimensions of rhythm itself—phase is the internal state of rhythm, and direction is the tendency of rhythm toward relational extension (see the entry on Direction in D.3 and Chapter 5). The triad $\mathcal{E}_i = (f_i, \theta_i, \mathbf{d}_i)$ is therefore not a juxtaposition of three independent properties, but a three-dimensional notation for the same concept of rhythm.

Computable Interface. The state variables are $\mathscr{E}_i = (f_i, \theta_i, \mathbf{d}_i)$ and relational weights $w_{ij}(\lambda)$; the dynamical interfaces are phase evolution (Appendix A:D1), weight updating (D2), cascade generation (D3), and the freezing criterion (D4). Output statistics may include global phase coherence, closure rate, freezing ratio, size of the strong-edge set, connectivity of frozen clusters, and so on.

Boundary of Misunderstanding. Not equivalent to continuous vibration: rhythm requires neither continuous trajectories nor sinusoidal form; its core lies in discrete evolution and settlement criteria. Not equivalent to subjective beat: the definition of rhythm does not depend on human sensation or time scales. Not equivalent to a single-frequency system: multiple frequencies may coexist; the key is whether closure and stable coupling structures form.

D.2. Completeness / Satisfaction

Standard Semantics. "Completeness" is the state in which an event or event cluster reaches settleable closure: its phase evolution achieves integer closure over some $\Delta\lambda$, and satisfies the stability threshold at the same scale, so that the structure can be retained as a "fact" and participate in subsequent generation.

Computable Interface. Closure condition (general form):

$$\theta_i(\lambda + \Delta\lambda) - \theta_i(\lambda) = 2\pi k_i, \quad k_i \in \mathbb{Z}.$$

Completeness is often used together with the freezing condition to distinguish "instantaneous closure" from closure that can be sustainably retained.

Boundary of Misunderstanding. Not equivalent to logical completeness: it has nothing to do with "completeness" in formal logic or set theory. Not teleological: it does not mean that the universe "pursues perfection"; it is a dynamical threshold. Not a one-time endpoint: completeness may appear repeatedly and emphasizes whether a certain settlement holds and is retainable.

Terminological Identity (Consistency across the Whole Book). In this book, "completeness" (Completeness / Satisfaction), "settled" (Settled, see D.11), and settlement in the criterion–bridging context refer to the same concept and may be used interchangeably. The three points of completeness—internality, termination, and recognizability—align item by item with the three requirements of the "genuine settlement criterion" (non-circularity, substantive nontriviality, and structural grounding; see Section 8.3 and the seven diagnostics in Appendix A, A.2). Whenever the reader encounters any one of these three terms in any chapter, it refers to the same threshold of existence: prior to readout, adjudicated by the generative process itself, and not constituted by anything external.

D.3. Direction

Standard Semantics. Direction is the tendency of rhythm toward relational extension—that is, the event's intrinsic tendency toward what kind of relation it extends—not the orientation of an event in external space. It exists at the ontological layer as a constitutive dimension of rhythm (together with phase as the two dimensions of rhythm), while the realization of extension belongs to the generation layer.

Computable Interface. Direction is represented by $\mathbf{d}_i$; however, what enters any generative adjudication is only directional difference—the relative orientation between two extension tendencies, represented by the directional dot product $\mathbf{d}_i \cdot \mathbf{d}_j$ (the directional kernel k_d, Appendix A:D2 and Section 9.3). The coordinate components of absolute direction $\mathbf{d}_i$ never enter any adjudication on their own. A global rotation changes no directional difference and therefore changes no adjudication (for formal verification, see Appendix B, B.12).

Boundary of Misunderstanding. Not equivalent to spatial orientation: direction does not point toward any spatial position or geometrical orientation, and it does not presuppose a space toward which it points; therefore it does not form a circle of "using direction to explain the nonexistence of geometry" (Section 5.3). Not equivalent to the direction of force: force belongs to the readout layer (Section 17.6), while direction belongs to the relational bias of the generation layer. Not an independent property parallel to rhythm: direction is a dimension of rhythm, and without rhythm it loses what it is (Section 5.4). Regarding fixity: direction is fixed before freezing and is not updated by phase dynamics (Section 5.3). The asymmetry in which phase updates at every step while direction is realized at every step but not updated is one structural source of the irreversibility of freezing (Sections 10.2 and 13.1). Regarding the source of propagation direction: propagation direction is not separately stipulated at the ontological layer, but emerges after the directed asymmetry of directional difference $k_d(i! \to !j) \neq k_d(j! \to !i)$ freezes (Section 9.3).

D.4. Difference

Standard Semantics. Difference refers to differences among events—rhythmic difference Δf_{ij}, phase difference $\Delta\theta_{ij}$, and directional difference (the relative orientation represented by $\mathbf{d}_i \cdot \mathbf{d}_j$). The ontological layer specifies the triad as the absolute and constitutive content of an event; the generation layer operates only through differences—absolute quantities cannot operate, while differences can (Section 5.5).

Key Properties. Difference is not an approximation method: it is not a numerical approximation under the continuous limit, but the only form that avoids background dependence (Section 20.5). Difference is the basis for "relation as operation on differences": relation is not an addition, because the generation layer processes only differences, and differences are essentially relations (Chapter 6). Difference is the mechanistic fulfillment of background independence:

because only differences enter adjudication, the global translation of absolute phase and global rotation of absolute direction cancel automatically within differences; therefore all generative adjudications are background-independent (for formal verification, see Appendix B, B.12).

Boundary of Misunderstanding. Not equivalent to a "small quantity": difference is not an infinitesimal in the differential sense, but a finite inter-event difference. Not equivalent to "measurement error": difference is a constitutive input of the generation layer, not an uncertainty at the readout layer.

D.5. Freezing

Standard Semantics. "Freezing" means that a closed event set enters a restricted generative state in which its key intrinsic variables no longer update freely as candidates, but participate in subsequent generation as constraints. Freezing is not stopping, but a role transformation from candidate to constraint (Sections 8.4 and 13.1).

Computable Interface. Rule constraint after freezing (general form): frozen events are removed from the generation-triggering set, but participate in subsequent constraints as acquired units; the frozen set is monotonically nondecreasing with λ, $\mathscr{F}(\lambda) \subseteq \mathscr{F}(\lambda+1)$ (Appendix A:C3).

Boundary of Misunderstanding. Not equivalent to rest: a frozen structure may still have internal updating; what it stops is "being regenerated as a candidate." Not equivalent to eternity: freezing is gated, and its qualification may be revoked in a hysteresis interval or reorganized in higher-level restructuring (the two legitimate meanings of "defreezing," Section 13.1). But once it has been inherited as an acquired predecessor by a higher layer, it is not reopened at the same layer.

D.6. Phase Simultaneity

Standard Semantics. Phase simultaneity means that a group of events completes phase-loop settlement within the same settlement window and is thereby adjudicated as "simultaneous" (Section 7.3). It is not coordinate simultaneity and does not depend on any global phase zero; by Lemma 1 of Appendix B, it is invariant under global phase translation.

Boundary of Misunderstanding. Not equivalent to "the same moment": simultaneity does not require sharing the same λ label, but only sharing the same settlement window. Not equivalent to global simultaneity: simultaneity is local, multiple, and capable of inconsistency; there is no unified "now" spanning the universe.

D.7. Event

Standard Semantics. An event is the basic unit of generation and carries a triadic intrinsic structure $(f, \theta, \mathbf{d})$. It is not a spacetime point, not a particle, not a process fragment, and not an observational result, but the bearer of generation–closure–freezing (Chapter 4). The three statuses of an event (possible, in generation, frozen) must not be collapsed.

Boundary of Misunderstanding. Not equivalent to "something that happens instantaneously": an event does not presuppose the time in which it happens. Not equivalent to a "point": an event is not embedded in any background space.

D.8. Relational Weight

Standard Semantics. Relational weight $w_{ij}(\lambda) \in [0,1]$ expresses the current generative reachability (qualification) of the relation between events i and j. Its updating is driven by the success or failure of closure, and all inputs are differences (Sections 6.6 and 9.3). Weight is not force and not probability, but the accounting of relational qualification.

Computable Interface. Projection updating (Appendix A:D2), compatibility kernel $K_{ij} = k_f \cdot k_\theta \cdot k_d$; weights can be reinforced, decayed, and cleared to zero; in general they may be directed ($w_{ij} \neq w_{ji}$).

D.9. Difference-Frequency Cascade

Standard Semantics. Difference-frequency cascade means that, when a new event is generated, the child-event frequency is given by the difference between parent-event frequencies (for example, $f_k = |f_i - f_j|$), accompanied by conservative frequency redistribution (Appendix A:D3 and A5). It is the minimal mechanism by which generation produces new structures under a conservation constraint.

D.10. Cosmic Logical Time, λ

Standard Semantics. $\lambda \in \mathbb{N}$ is the count of generative depth, not a time parameter. It is not continuous, not uniform, and not reversible; it records only the number of steps in which generative rules have been successfully applied and settlement completed (Section 7.2). It retains the sequential intuitions of partial-order reachability and cumulative depth, while discarding the temporal intuitions of uniform passage, continuous instantaneity, global simultaneity, and reversibility.

Boundary of Misunderstanding. Not equivalent to "discrete time": discrete time is still time, while λ does not belong to the family of temporal concepts. Not equivalent to a "clock reading": λ has no unit, and observable time is its readout through some clock.

D.11. Settlement / Closure

Standard Semantics. "Settlement" means a completion event in which the closure condition holds over some $\Delta\lambda$ and produces a retainable constraint consequence. Settlement is the operational criterion of "fact generation," transforming "fact" from a narrative statement into an adjudicative condition: when something counts as completed, and when it is not retained.

Computable Interface. The input is an event set and its trajectory segment; the outputs are whether a settlement window exists, the minimal $\Delta\lambda$, and post-settlement stability indicators. Settlement is often used together with the freezing criterion to distinguish "closed but not sustainable" from "closed and retainable."

Boundary of Misunderstanding. Not equivalent to a synchronization process: synchronization is an approach dynamic, while settlement is an adjudication event. Not equivalent to measurement collapse: settlement requires no external observation to be triggered.

Identity with Completeness. "Settlement" and D.2 "Completeness" refer to the same concept (see D.2, Terminological Identity). As a "genuine settlement predicate," settlement must pass the seven diagnostic tests (non-circularity, substantive nontriviality, structural grounding, settled predecessor prior to readout, avoidance of constitutive inversion, avoidance of closed internality, and nonaccidental downstream bearing). Its formalization is given in Appendix A, A.2. These seven items ensure that settlement is a legitimate criterion prior to readout and not constituted by readout, rather than an accidental pause in a process or a pseudo-completion retrospectively recognized by observation.

D.12. Readout Protocol

Standard Semantics. Readout protocol refers to the set of limitations followed by any finite-resolution physical process when sampling frozen structures. It consists of the triple truncation of bandwidth, gating, and threshold (Chapters 14 and 18). A readout protocol is not an observer and does not presuppose consciousness or macroscopic apparatus—any finite-resolution physical process performs readout.

Computable Interface. Generation-layer operator counterparts of the triple truncation: temporal granularity $\leftarrow 1/f_{\max}$, spectral boundary $\leftarrow f_{\max}$, triggering threshold $\leftarrow$ freezing criterion.

D.13. Emergence

Standard Semantics. Emergence means that readout-layer structures (time, space, causality, probability, force, field) appear as lossy readouts of frozen results at the generation layer. Emergence is not the generation of new entities at the generation layer, but the organizational appearance of generative facts at the readout layer (Chapters 15–18 and 22).

Boundary of Misunderstanding. Not equivalent to "new properties of complex systems": emergence in this book is layer emergence (readout layer versus generation layer), not same-layer complexity emergence.

D.14. Phase Coupling

Standard Semantics. Phase coupling refers to whether the phase configuration of multiple events satisfies the same closure criterion, that is, generative selection rather than interaction (Section 8.2). It transmits no entity, requires no propagation speed, and has phase difference as its only input.

Computable Interface. The coupling term in phase updating, such as $\sum_j w_{ij}\sin(\theta_j - \theta_i)$.

D.15. Stability Domain / Freezing Basin

Standard Semantics. A stability domain refers to a relational substructure in the weight network whose relations mutually reinforce one another and possess recovery capacity under perturbation (Section 9.2). It is not a potential well in phase space, but a history in relations; its "depth" is the historical length of freezing and the defreezing threshold, rather than the low point of an energy surface.

Boundary of Misunderstanding. Not equivalent to a potential well: a stability domain is not located in any presupposed space and corresponds to no minimum of any potential function. Not equivalent to equilibrium: it is a relational organization of repeated success, not a balance of forces or energy.

E. Numerical Settings and Computational Details of the Y-Type Four-Event Condensate

This appendix gives a reproducible numerical implementation of the Y-type four-event condensate: from variables, normalization, and unit mapping (E.1), to the discrete implementation of dynamical rules (E.2), initialization of the Y-type structure (E.3), numerical adjudication of co-presence settlement (E.4), freezing criteria and stability measures (E.5), the main calculation of the mass difference (E.6), readout of mass and mass difference (E.7), typical parameter sets and robustness checks (E.8), computational workflow (E.9), and result-presentation template (E.10). This appendix follows the same discipline as Appendix C: absolute mass is display calibration, while the mass difference Δm bears predictive responsibility, and Δm is anchor-independent.

E.1. Variables, Normalization, and Unit Mapping

E.1.1. Event State Variables

Each event i carries $(f_i, \theta_i, \mathbf{d}_i)$, and relational weights $w_{ij}(\lambda) \in [0, 1]$ are carried between events. The system evolves over discrete logical time λ.

E.1.2. Readout Mapping from Normalized Frequency to MeV

The unit mapping takes $m \propto \varepsilon_0 \cdot f$, where ε_0 is the energy scale (MeV/frequency unit). In the demonstration configuration one may take $\varepsilon_0 = 1$, that is, directly report f_i on the MeV scale. It must be emphasized that this choice only fixes the display scale and does not constitute a dynamical input. For strict alignment, ε_0 should be calibrated separately as part of the readout protocol and kept consistent across different observables. This is consistent with the mass discipline in Section 16.2—unit mapping belongs to external calibration and bears no predictive responsibility.

E.2. Discrete Implementation of the Dynamical Rules

E.2.1. Phase Evolution (Coupled Phase Updating)

Phase is updated according to D1 (Appendix A):

$$\theta_i(\lambda+1) = \mathrm{wrap}\left(\theta_i(\lambda) + 2\pi f_i \Delta\lambda + \alpha \sum_{j\in N(i)} w_{ij}\sin(\theta_j - \theta_i)\right),$$

with $\Delta\lambda \equiv 1$, and α as the synchronization coupling strength. The coupling term reads only the phase difference $\theta_j - \theta_i$, so phase evolution is background-independent (Appendix B, Lemma 1).

E.2.2. Weight Updating (Compatibility-Driven Reinforcement)

Weights are updated by the projection update of D2 (Appendix A), with compatibility kernel

$$K_{ij} = k_f(\Delta f_{ij}), k_\theta(\Delta\theta_{ij}), k_d(\mathbf{d}_i \cdot \mathbf{d}_j),$$

where the three factors read only rhythmic difference, phase difference, and directional difference respectively. Weights are kept bounded within $[0,1]$ through projection.

E.3. Initialization of the Y-Type Structure

E.3.1. Direction Initialization (Three Directions at 120° in the Plane)

The directions of the three peripheral events are initialized with equal pairwise relative orientation. In coordinate bookkeeping, this appears as a 120° configuration in the plane, with vector sum equal to zero. This "120°" must be understood according to the welding statement of Appendix B and Section 19.2: it is bookkeeping for the balanced relative orientations of three relational extension tendencies, can be globally rotated without changing any directional difference or adjudication, and does not presuppose angular position in space.

E.3.2. Frequency Initialization (Center–Periphery Scale Separation)

Take

$$f_c = 930, \qquad f_{qk} = 300 \quad (k = 1,2,3),$$

and allow small adaptive corrections to the peripheral frequencies during evolution (see E.6), used to implement the "tension–compensation" closed loop. It must be emphasized that $f_c = 930$ is only a display anchor (used to report directly on the MeV scale), is not a predicted quantity, and does not constitute a falsifiable dynamical input.

E.3.3. Phase Initialization (In-Phase Configuration vs Opposite-Phase Configuration)

Flavor is distinguished by peripheral phase labels: the up type corresponds to $\theta = 0$, while the down type corresponds to $\theta = \pi$. The proton-like configuration (uud) has peripheral phases $(0, 0, \pi)$, while the neutron-like configuration (udd) has $(0, \pi, \pi)$.

E.3.4. Weight Initialization

Center–periphery weights and periphery–periphery weights are initialized according to the Y-type skeleton, with specific value ranges given in E.8.

E.4. Numerical Adjudication of Co-Presence Settlement

E.4.1. Closure Criterion (Approximation of Integerness)

Closure is adjudicated by an integerness approximation of phase-loop closure: the sum of loop phase differences falls within the tolerance $\left|\sum \Delta\theta_{ij}\right| \leq \varepsilon_\theta$.

E.4.2. Principle for Choosing the Settlement Window

Settlement requires closure to hold repeatedly within a finite window $\Delta\lambda$, and not to depend on the internal ordering of steps within the window (Sections 7.3 and 20.4), rather than being accidentally satisfied in a single step. The minimum window length is determined by the number of steps required for repeated confirmation.

E.5. Freezing Criteria and Stability Measures

E.5.1. Aggregation Strength (Connectivity)

The connectivity of the frozen cluster is used to measure structural aggregation strength.

E.5.2. Phase Divergence (Drift Strength)

Phase divergence γ_i is used to measure phase-drift strength—this quantity is also the basis for adjudicating the sign of the central frequency shift in E.6.2.

E.5.3. Freezing Criterion (Reproducible Simplified Version)

An event freezes if and only if three kinds of conditions are simultaneously satisfied: phase locking (aggregation strength reaches the threshold), phase divergence lies within tolerance, and directional stability holds.

E.6. Main Calculation of the Mass Difference: Flavor-Difference Term and Central Frequency-Shift Term

E.6.1. Dominant Term—Numerical Setting of the Flavor Difference $(f_d - f_u)$

The dominant contribution to the mass difference comes from the frequency difference between peripheral up-type and down-type events, determined by the flavor phase labels (E.3.3). In the numerical implementation, up-type and down-type peripheral events are assigned slightly different frequency scales f_u, f_d, and their difference $f_d - f_u$ is an input difference determined by the configuration and requiring no fitting. Under the original calibration, $f_d - f_u = 3.13 \times 10^{20}$ Hz (corresponding to ≈ 1.29 MeV on the reporting scale). The neutron-like configuration $(0, \pi, \pi)$ has one more down-type periphery than the proton-like configuration $(0, 0, \pi)$, and the flavor-difference term therefore makes a positive contribution to Δm. This term does not read the central frequency f_c and is therefore anchor-independent.

E.6.2. Correction Term—Numerical Treatment of the Central Frequency Shift $\Delta \bar{f}_g$

The two $\theta = \pi$ peripheries of the neutron-like configuration introduce greater phase tension (measured by the phase divergence γ_i in E.5.2). Through nonlinear feedback, this tension slightly raises the central event's freezing frequency, so $\Delta \bar{f}_g > 0$. In the numerical implementation, the sign of $\Delta \bar{f}_g$ is determined by the configuration difference in γ_i (larger neutron-like $\sum \gamma_i \to \Delta \bar{f}_g > 0$), but its specific numerical value is treated as an estimate (in the original source, $\Delta \bar{f}_g \approx 0.5 \times 10^{20}$ Hz ≈ 0.21 MeV). Honest marking: an attempt was made to derive $\Delta \bar{f}_g$ endogenously from the micro-adjustment loss in E.6.3, but under the minimal model this derivation was numerically unstable (divergent). It is therefore treated as an estimated small correction and entered separately into the error budget (aligned with

Appendix C.3.3).

E.6.3. Secondary Effect—Peripheral Frequency Micro-Adjustment δf

Peripheral frequency micro-adjustment

$$f_{qk}(\lambda+1) = f_{qk}(\lambda) - \eta, \partial\Phi/\partial f_{qk}$$

(with the loss Φ as in the main text) is retained as a secondary effect, but no longer serves as the dominant term of Δm. The reasons (aligned with Appendix C.3.5) are as follows: this δf is obtained through the matching kernel and threshold function, is sensitive to the central anchor f_c, and its sign is unstable under the minimal model. Therefore it cannot bear the main body or sign of Δm; it serves only as a secondary response of the periphery to opposite-phase tension, is counted in the error budget, and does not enter the dominant term of Δm.

E.7. Readout of Mass and Mass Difference

E.7.1. Absolute-Mass Readout

After freezing, the mass scale is estimated by aggregating the frequency budget, $m \propto \sum_i f_i$. $\varepsilon_0 = 1$ is only a display scale and does not constitute a dynamical input. About 98% of absolute mass comes from the central event f_c (consistent with QCD baryon mass being gluon-dominated). Therefore absolute mass is a consistency check under display calibration and bears no predictive responsibility.

E.7.2. Calculation Path of the Mass Difference

According to the main mechanism in E.6, the mass difference is

$$\Delta m = \frac{h}{c^2}\left[(f_d - f_u) + \Delta \bar{f}_g + O(\sigma_f^2)\right],$$

that is, it consists of the flavor-difference dominant term (parameter-free, ≈ 1.29 MeV, anchor-independent) plus the central frequency-shift correction term (estimated, ≈ 0.21 MeV, with its sign determined by the γ configuration difference). The total is rounded according to the original source to ≈ 1.3 MeV, with positive sign (the neutron-like configuration is heavier).

Anchor Independence. Both terms in Δm are frequency differences, and the central anchor f_c does not enter. Therefore, in the numerical implementation, when f_c is taken as 600/800/930/1100/1300/1500 (on the MeV reporting scale) while all else remains unchanged, absolute mass translates with f_c, whereas Δm remains constantly ≈ 1.3 MeV (with relative change zero within numerical precision). The full sensitivity table is given in Section 20.5. This contrasts with the old "$\Delta m \approx 2\delta f$" version—the old δf was obtained through the matching kernel $g(f_c, f_i)$ and strongly depended on f_c (numerical tests show that, when f_c ranges from 600 to 1500, Δm fluctuates by hundreds of times and even changes sign), which is a problematic simplification. The new main mechanism no longer has this problem.

E.8. Typical Parameter Sets and Robustness Checks

E.8.1. A Reproducible Demonstration Parameter Set (Ranges)

Give reproducible value ranges for coupling strength α, weight-update rate, freezing thresholds, tolerance ε_θ, and so on, together with the initialization settings of E.3. Specific numerical values are subject to the reproducible script.

E.8.2. Robustness Checks—Replacing the Doubt about Sign Robustness with Anchor Independence

The minimum reporting set is as follows: (1) anchor sensitivity: when $f_c \in [600, 1500]$ varies, Δm should remain unchanged at ≈ 1.3 MeV (this is the direct test of anchor independence in the main mechanism and replaces the old doubt about "whether the sign is stable"—the old version was sign-unstable precisely because of $2\delta f$); (2) flavor-difference independence: whether the main body of Δm varies linearly with $(f_d - f_u)$ and does not vary with f_c; (3) sign of the central correction: whether neutron-like $\sum \gamma_i$ is systematically larger than proton-like $\sum \gamma_i$ (ensuring $\Delta \bar{f}_g > 0$); (4) directional perturbation test: when the peripheries deviate from 120°, the freezing probability declines, testing whether the Y-type is an attractor rather than being hard-coded.

E.9. Computational Workflow (Workflow Level Reproduction Protocol)

Initialize the Y-type structure (E.3) $\rightarrow$ iterate phase evolution and weight updating (E.2) $\rightarrow$ adjudicate co-presence settlement (E.4) $\rightarrow$ adjudicate freezing and stability (E.5) $\rightarrow$ read out absolute mass (E.7.1, consistency check under calibration scale) $\rightarrow$ calculate the mass difference (E.7.2, by flavor-difference dominant term + central estimated correction) $\rightarrow$ scan the f_c anchor to verify the invariance of Δm. The full workflow is reproducible in parameters and script.

E.10. Result-Presentation Template

Reported items include: f_c and peripheral frequencies at freezing, connectivity and phase divergence of the frozen cluster, absolute-mass readout (marked as a consistency check under display calibration), decomposition of the mass difference Δm into the flavor-difference dominant term and central correction term, and Δm under the f_c anchor scan (verifying constancy).

E.11. Appendix Summary

This appendix gives a reproducible numerical implementation of the Y-type four-event condensate: phase locking $\rightarrow$ freezing criterion $\rightarrow$ mass readout. On the mass issue, this appendix follows the same discipline as Appendix C: absolute mass (about 98% from the central anchor f_c) is a consistency check under display calibration and bears no predictive responsibility; what bears falsifiable responsibility is the mass difference Δm, which consists of the parameter-free flavor-difference dominant term (≈ 1.29 MeV) plus the central frequency-shift correction term with determined sign and estimated value (≈ 0.21 MeV), giving a total of ≈ 1.3 MeV with the correct sign and anchor independence. The old simplified version "$\Delta m \approx 2\delta f$" has been downgraded to a secondary effect counted in the error budget because it depends on the central anchor and has an unstable sign. Absolute mass translates with the anchor while Δm remains unchanged, providing numerical evidence for the separation between calibration and prediction.

F. Numerical Settings and Computational Details of the Four-Event Proton Steady State

This appendix provides a reproducible numerical implementation of the four-event proton-like steady state, complementing Appendix E (the general numerical implementation of the Y-type condensate). Appendix E focuses on the general dynamics of the Y-type structure and the calculation of mass differences, whereas Appendix F focuses on phase locking, freezing, absolute-mass readout of a single proton-like steady state, and comparative outputs of charge and spin proxies. This appendix follows the same discipline as Appendices C and E: absolute mass is a consistency check under display calibration, while the mass difference Δm is what bears predictive responsibility.

F.1. Event State Variables and Conventions

The proton-like steady state consists of four events: one central event and three peripheral events. Each event carries $(f_i, \theta_i, \mathbf{d}_i)$, and relational weights $w_{ij}(\lambda) \in [0,1]$ are carried between events. The dynamical rules use only the relative magnitudes of f_i, the comparison structure of the matching kernel and thresholds, and do not use any absolute dimensional quantity (as stipulated in F.1.2).

F.2. Initial Configuration and Frequency Settings

F.2.1. Frequency Settings (Prior Statement on the Display Anchor)

The central event is set to $f_c = 930$ (on the MeV reporting scale), and the three peripheral events are set to approximately equal lower frequency scales. At this point, a statement that runs throughout this appendix must be made in advance, to prevent the absolute-mass values below from being misread.

Statement on the display anchor (important). Here $f_c = 930$ (on the MeV reporting scale) is a display energy-scale anchor, used to report the central frequency directly on the MeV scale and to facilitate comparison with the order of magnitude of hadronic masses. It is not a predicted quantity and does not constitute any dynamical input—the dynamical rules use only the relative magnitudes of f_i, the matching kernel, and the comparison structure of thresholds (as stipulated in F.1.2). Therefore, any absolute-mass value read out from it (such as m_p in F.5.3) is a calibration product rather than a prediction: absolute mass depends on the externally calibrated κ and m_0 in the external readout mapping $m = \kappa\mathfrak{m} + m_0$. What bears the falsifiable responsibility of this appendix is not absolute mass, but the mass difference Δm (see F.5.4, Appendix C, and Section 20.5). In the difference, the zero point is eliminated and f_c is not read; it is therefore anchor-independent. This statement prevents numerical values such as 938.3 below from being misread as a fit in which the input is chosen to match the output.

F.2.2. Phase and Direction Initialization

The flavor of peripheral events is distinguished by phase labels (the up type corresponds to $\theta = 0$, and the down type corresponds to $\theta = \pi$). The proton-like configuration (uud) has peripheral phases $(0, 0, \pi)$. The directions of the three peripheral events are initialized with balanced relative orientations (a planar 120° configuration in coordinate bookkeeping; for its relation-only meaning, see Appendix B and Section 19.2).

F.3. Phase Evolution and Weight Updating

Phase is updated according to the coupled phase update rule (D1 in Appendix A):

$$\theta_i(\lambda+1) = \mathrm{wrap}(\theta_i(\lambda) + 2\pi f_i \Delta\lambda + \alpha \sum_j w_{ij} \sin(\theta_j - \theta_i)),$$

where the coupling term reads only phase differences. Weights are updated by compatibility-driven projection updating (D2). The compatibility kernel $K_{ij} = k_f, k_\theta, k_d$ reads rhythmic difference, phase difference, and directional difference respectively, and weights are bounded within $[0,1]$.

F.4. Steady State and Freezing Criteria

The proton-like steady state is reached if and only if the four events repeatedly satisfy the phase-loop closure condition within a finite settlement window and pass the freezing criteria (phase locking, phase divergence within tolerance, and directional stability). The phase-locked structure of the steady state gives converged weights w_{ij} and a stable distribution of phase differences, the latter serving as the input for calculating the correction term in F.5.

F.5. Absolute-Mass Readout and Mass Difference

F.5.1. Structure of the Mass Readout

After freezing, the mass scale is estimated by aggregating the frequency budget. Approximately 98% of the absolute mass comes from the central event f_c, and the remainder comes from peripheral contributions and small corrections induced by phase tension (consistent with the QCD view that baryon mass is gluon-dominated).

F.5.2. Source of the Correction Term

In the phase-locked steady state, the correction term δm_{corr} induced by the peripheries and tension is calculated from the converged w_{ij} and phase differences.

F.5.3. Baseline Numerical Check—938.3 as a Consistency Check under the Display Scale

The baseline setting takes $f_c = 930$ (the display energy-scale anchor; see the statement in F.2.1). In the phase-locked steady state, the correction term is calculated from the converged w_{ij} and phase differences, yielding

$$m_p^{(\text{GOCD})} \approx 930 + 8.3 = 938.3 \quad (\text{MeV, display scale}).$$

The status of this value must be made explicit: it is a consistency check under the display anchor $f_c = 930$, used to display the structural decomposition of "central dominance (f_c, about 98%) + small correction ($\delta m_{\text{corr}} \approx 8.3$)." It is not an independent prediction. The closeness of ...938.3 to the experimental value 938.272 should not be interpreted as the model matching experiment—because the 930 in 938.3 is a specified display anchor, and 8.3 is a correction term. If the anchor is instead set to $f_c = 600$, the same structure gives

$\approx$ 608.3, and so on (for the anchor scan, see Section 20.5). In other words, absolute mass translates with the display anchor. It tests the self-consistency of the "dominant term + correction" decomposition, but not any predicted absolute value. Accordingly, this appendix does not claim the closeness of 938.3 to experiment as evidence of model success; the falsifiable mass-related claim of the model is concentrated in Δm in the next subsection.

F.5.4. Mass Difference Δm and Anchor Independence

F.1–F.5.3 of this appendix calculate the absolute mass of the proton-like steady state. It depends on the display anchor f_c, is a calibration product, and bears no predictive responsibility. What bears the falsifiable responsibility of this framework on the mass problem is the proton–neutron mass difference Δm. Its full mechanism is given in Appendix C.3.3, its numerical settings in Appendix E.6–E.7, and its sensitivity analysis in Section 20.5. The key points are restated here to ensure the self-consistency of this appendix.

Δm is given by the two terms in the original source's Eq.3—the flavor-difference dominant term $(f_d - f_u)$ (parameter-free, $\approx$ 1.29 MeV, determined by the peripheral *uud* versus *udd* configurations, and anchor-independent) plus the central frequency-shift correction term $\Delta \bar{f}_g$ (estimated, $\approx$ 0.21 MeV, with its sign determined by the phase-tension configuration difference). Together they give $\approx$ 1.3 MeV, with positive sign (the neutron-like configuration is heavier). The key point is that both terms in Δm are frequency differences, and the central anchor f_c does not enter. Therefore, when f_c is set to 600/800/930/1100/1300/1500 (MeV) while all else remains unchanged, the absolute-mass readout in this appendix (F.5.3) translates with f_c ($\approx 608/808/938/1108/1308/1508$), whereas Δm remains constantly $\approx$ 1.3 MeV. This contrast—absolute mass translating with the anchor while Δm remains unchanged—is precisely the numerical evidence for the separation between calibration and prediction. It also transforms the objection that $f_c = 930$ is "fitting numbers to numbers" into supporting evidence: what can be freely replaced without affecting the prediction is a display anchor, not a fitted parameter (for the full table, see Section 20.5).

F.6. Charge and Spin Proxies

Charge and spin are read out through configuration proxy quantities: the proton-like configuration gives charge proxy $Q = +1$ and the corresponding spin proxy, while the neutron-like configuration gives $Q = 0$. A positioning sentence must be added here: calibration terms in the charge proxy (such as the background density term $\rho_{\text{bg}}V$ and the normalization $\mathcal{N}_Q$) belong to readout-layer calibration, just like κ, m_0 in absolute mass. The robust object of testing is the discrete output change between comparative configurations (the difference between proton-like $Q = +1$ and neutron-like $Q = 0$), rather than the absolute assignment—consistent with the mass discipline that differences take priority over absolute values.

F.7. Appendix Summary

This appendix gives a reproducible calculation of the four-event proton-like steady state: phase locking $\to$ freezing criterion $\to$ the "dominant term + correction" decomposition of absolute-mass readout. It must be emphasized that the absolute mass (938.3) is a consistency check under the display anchor $f_c = 930$, not a prediction. What bears falsifiable responsibility is the mass difference Δm (flavor-difference dominated, anchor-independent, ≈ 1.3 MeV; see F.5.4 and Appendices C and E), which is the predictive content of this framework on the mass problem. Absolute mass translates with the anchor, while Δm remains unchanged, providing numerical evidence for the separation between calibration and prediction.

G. A Candidate Joint-Closure Construction for Bell/CHSH Correlations

— A conditional, fallible, readout-layer formulation without ontic probability

This appendix studies a precisely delimited question: can GOCD, under the constraints of its own principles, organize quantum-type Bell/CHSH correlations into a joint-closure structure that is globally nonseparable, operationally no-signaling, without postselection, and without ontic probability? It does not attempt to make a Bell-local hidden-variable model violate CHSH—that has already been excluded by Bell's theorem and repeatedly confirmed by the naive numerical attempts previously made in this appendix (G.3). It studies a different question: if entangled readout is not two local objects each carrying a pre-stored result, but rather the joint freezing of the same shared generative process at two readout interfaces, can such a structure formally accommodate both quantum-type correlations and operational no-signaling?

The status of this appendix is strictly conditional, and this must be stated at the outset to avoid misreading it as a completed derivation. This appendix will show that, under a set of explicitly listed specialized assumptions, one can write the joint-closure capacity

$$K_{rs}(a,b) = \tfrac{1}{4}\left[1 - rs\cos 2(a-b)\right],$$

and from it derive $E(a,b) = -\cos 2(a-b)$ and $|S| = 2\sqrt{2}$, while preserving operational no-signaling. But this appendix does not claim that this capacity form has already been derived from GOCD's discrete phase–weight–direction dynamics. On the contrary, this appendix will honestly point out (G.8) that the cosine form is a result-level assumption, not a conclusion uniquely derived from weaker structural assumptions (such as "finite additivity")—a counterexample shows that the same set of weak assumptions also permits a local model that yields only $|S| = 2$. Therefore, this appendix provides a candidate construction that is openly assumed, computable, and capable of failure. It marks precisely which step remains an assumption and which step is actually derived, while listing "generating this capacity from the foundational dynamics" as an unfinished constructive task.

G.1. The Status of This Appendix: From an "Unfinished Toy Model" to a "Candidate Construction"

In previous formulations, Appendix G was an unfinished numerical task: to test whether the discrete phase mechanism of GOCD could produce CHSH violation. Numerical results from naive shared-phase models showed that, as long as the readout can still be written as two local functions $A = A(a, \lambda)$ and $B = B(b, \lambda)$, and the distribution of λ does not depend on the measurement settings, the model stays close to the local bound $|S| \approx 2$ and does not violate CHSH. This is not an accidental numerical difficulty, but the direct expectation of Bell's theorem: any implementation satisfying Bell-local factorization, no matter how many phase, direction, or weight variables are added, necessarily belongs to the class covered by Bell's theorem, as long as it can ultimately still be written as two local readout functions, and therefore must satisfy $|S| \leq 2$.

This appendix therefore no longer adjusts parameters along the route of "local shared phase + local deterministic readout"—that route is closed by the theorem. Instead, it turns to another readout-generation structure:

$$(A,B) = \mathscr{C}(\Lambda; a, b),$$

where Λ is the finite shared generative domain of the entangled relation in this run, a, b are Alice's and Bob's local gates, and (A, B) are the two local readouts jointly frozen after whole closure. This structure is explicitly different from $A = A(a, \lambda)$ and $B = B(b, \lambda)$: in the GOCD understanding, an entangled pair is not two local entities each carrying pre-stored results, but the joint freezing of the same generative process at two readout interfaces. The result finally appears as two local facts, but they are not the simple product of two independent local generative processes.

Thus, the core question studied in this appendix is:

$$P(A,B \mid a,b,\Lambda) \neq P(A \mid a,\Lambda)P(B \mid b,\Lambda)$$

can hold simultaneously with the operational no-signaling conditions

$$P(A \mid a,b) = P(A \mid a), \qquad P(B \mid a,b) = P(B \mid b).$$

That is, the goal is not to construct a Bell-local model that violates Bell's theorem, but to construct a non-Bell-local factorizable, yet operationally no-signaling, joint-closure model—precisely the structure quantum mechanics itself possesses.

G.2. Strict Layering of Three Classes of Assumptions

GOCD is a general generative framework, not a closed set of equations that has already uniquely determined all phenomena. Introducing specialized assumptions in a concrete problem is not a defect, but a necessary step through which the framework obtains physical content. However, to avoid "secretly assuming the quantum answer and then claiming to have derived it," this appendix strictly divides all assumptions into three classes and requires every step of the derivation to mark which class it uses.

First class: GOCD universal constraints. No a priori spacetime, no ontic probability, generation by differences, finite-step closure, and irreversible freezing. Any model that introduces a pre-stored spacetime background, ontic random sampling, completion by an infinite limit, or unfailable closure is no longer a GOCD implementation.

Second class: specialized assumptions for the Bell problem (structural). These are structural choices made when applying the framework to entangled readout: joint-closure form $(A,B) = \mathscr{C}(\Lambda;a,b)$, axial equivalence of polarization gating $a \equiv a+\pi$, mutual exclusivity and completeness of the four joint freezing channels, opposite-sign freezing of an anticorrelated source under identical gates, and single-side capacity symmetry. These are legitimate structural assumptions, but they do not uniquely determine the specific functional form of the correlation (see G.8).

Third class: result-level assumptions. These are assumptions that directly write in, or are equivalent to, the target result, such as directly stipulating the cosine form of the capacity $K_{rs} = \frac{1}{4}(1 - rs\cos 2(a - b))$, or the correlation $E = -\cos 2(a-b)$. Result-level assumptions may be used to establish an effective description, but their explanatory power is weak, because the target result has already been presupposed.

The honesty of this appendix lies in showing that the cosine form of K_{rs} belongs to the third class rather than the second class—even though G.8 previously attempted to argue, through "finite additivity," that it was a consequence of the second class, that argument does not hold. Making this classification explicit is the key difference between the present rewrite and the earlier version.

G.3. Why the Naive Local Shared-Phase Model Is Insufficient (Numerical and Structural Confirmation)

Let the source generate a shared phase variable λ, with local readouts $A = \operatorname{sign}\cos(\theta_A(\lambda) - a)$ and $B = \operatorname{sign}\cos(\theta_B(\lambda) - b)$. As long as this can be written as $A = A(a,\lambda)$ and $B = B(b,\lambda)$, with the distribution of λ independent of (a,b), Bell-local factorization $P(A,B \mid a,b,\lambda) = P(A \mid a,\lambda)P(B \mid b,\lambda)$ holds, and one must have $|S| \leq 2$.

This point has been confirmed by multiple independent numerical experiments: strict reruns of such naive implementations (including variants with symmetric cross-coupling, decay, joint angular modulation, and dynamic weights) always give $|S|$ equal to 2.0 within numerical precision, and never cross it. Its structural root can be verified in one line: in these implementations, Alice's readout A has zero dependence on Bob's angle b at the sample level, and therefore A is indeed a local function of (λ, a). Thus, its failure is not due to poor parameter tuning, but is structural: any implementation compressed into two local response functions inevitably falls back into the Bell-local class. This shows that, if GOCD is to handle Bell correlations, the key is not to add local variables, but to change the readout-generation structure—the two readouts must appear as the joint freezing results of one whole closure process.

G.4. The Joint Closure Domain Λ

Denote the finite shared domain of one entangled generation by

$$\Lambda_{AB} = \left(\mathscr{E}_A, \mathscr{E}_B, \mathscr{G}_{AB}, \Delta\theta_{AB}, \Delta\mathbf{d}_{AB}, w_{AB}, h_{AB}\right),$$

where $\mathscr{E}_A, \mathscr{E}_B$ are the event structures at the two readout interfaces, $\mathscr{G}_{AB}$ is the shared relational structure that has not yet been decomposed into two local facts, $\Delta\theta_{AB}$ and $\Delta\mathbf{d}_{AB}$ are the relational phase difference and directional difference, w_{AB} is the shared relational weight, and h_{AB} is the generative history of this run. Λ_{AB} is not the state of the whole universe, nor a pre-stored hidden-variable sample space; it is only the finite generative domain in which this entangled event can participate in closure, and before freezing it cannot be decomposed into two completed local facts.

When readout occurs, gates a, b are not external background coordinates, but relational boundaries imposed by two frozen readout devices on the shared generative domain. They enter the same joint closure process:

$$\Lambda_{\lambda+1} = \mathscr{U}_{\mathrm{GOCD}}\big(\Lambda_\lambda; G_A(a), G_B(b)\big),$$

where λ is the discrete generative step. If joint closure completes within a finite step λ_*, the freezing map gives $(A,B) = \mathscr{F}(\Lambda_{\lambda_*}; a, b)$. If closure cannot complete within a finite window, that generation fails, and completion cannot be replaced by infinite approximation or "not yet converged."

G.5. Axial Relational Representation of Polarization Gates (This Step Is a Legitimate Derivation)

For polarization-type Bell experiments, the measurement setting is an unoriented axis, so $a \equiv a + \pi$. The ordinary directed representation $(\cos a, \sin a)$ changes sign under $a \to a + \pi$, and therefore cannot represent the same polarization axis. The minimal relational representation satisfying axial equivalence is

$$\mathbf{q}(a) = (\cos 2a, \sin 2a), \qquad \mathbf{q}(a+\pi) = \mathbf{q}(a).$$

The only minimal axial relational scalar between two gates is

$$\mu(a,b) = \mathbf{q}(a)\cdot\mathbf{q}(b) = \cos 2a \cos 2b + \sin 2a \sin 2b = \cos 2(a-b).$$

This step is a legitimate structural derivation (a direct consequence of the second-class assumptions): it uses only the fact that, at freezing, a polarization readout device can distinguish axial relations but not the arrow direction of the axis. $\mathbf{q}(a)$ is the relational representation of the readout gate, not a spatial direction presupposed at the generation layer. It must be emphasized that this step gives only that "correlation can enter only through $\mu = \cos 2(a-b)$." It does not stipulate how the correlation depends on μ; the latter is precisely the real gap identified in G.8.

G.6. The Four Mutually Exclusive Joint Freezing Channels (Structural Assumptions)

Each side has binary readout $A, B \in \{-1, +1\}$. The joint results are four types: $(++), (+-), (-+), (--)$, corresponding to closure channels $\mathscr{C}_{rs}$. This candidate construction makes the following structural assumptions about them (second class): mutual exclusivity (one successful closure freezes into only one channel), completeness (the result of successful closure must belong to one of the four channels), single-side symmetry ($\sum_s K_{rs} = \sum_r K_{rs} = \frac{1}{2}$, ensuring no-signaling), and anticorrelated-source closure (when $a = b$, $K_{++} = K_{--} = 0$, $K_{+-} = K_{-+} = \frac{1}{2}$). These assumptions are legitimate and natural, but as G.8 will show, they are insufficient to uniquely determine the dependence of capacity on μ.

G.7. Closure Capacity: The Cosine Form and Its Algebraic Consequences (This Is a Result-Level Assumption)

Let $K_{rs}(a,b)$ denote the closure-channel capacity that can freeze into the joint result (r,s) under a fixed finite readout protocol. It is not the ontic probability of a single event randomly selecting the result (r,s), but the capacity of the actual generative relation, under given gates, to complete that class of joint closure. This candidate construction adopts the following capacity form:

$$K_{rs}(a,b) = \tfrac{1}{4}\big[1 - rs\cos 2(a-b)\big] = \tfrac{1}{4}(1 - rs\mu).$$

It has all required properties (each can be verified directly, and has been numerically checked): completeness $\sum_{r,s} K_{rs} = 1$; nonnegativity $K_{rs} \geq 0$ (since $\mu \in [-1,1]$); single-side symmetry $\sum_s K_{rs} = \sum_r K_{rs} = \frac{1}{2}$; when $a = b$, $\mu = 1$, yielding perfect anticorrelation $K_{++} = K_{--} = 0$, $K_{+-} = K_{-+} = \frac{1}{2}$; and when $a - b = \frac{\pi}{4}$, $\mu = 0$, the four channels have equal capacity $\frac{1}{4}$.

The epistemic status of this capacity form must be marked explicitly: it is a third-class (result-level) assumption in the sense of G.2, not a conclusion derived from the first- and second-class assumptions. It satisfies all the consistency requirements above, but capacity forms satisfying these requirements are not unique (see G.8). This appendix adopts the cosine form because it corresponds to the known quantum correlation; adopting it is equivalent to assuming the result of the quantum Born rule in this case. This appendix does not conceal that point—this is precisely where it differs from a completed derivation.

G.8. Key Clarification: Why the Cosine Form Cannot Be Derived from "Finite Additivity"

The previous argument in this appendix once attempted to promote the cosine form to a consequence of second-class (structural) assumptions. The reasoning was as follows: the closure capacity should be finitely additive over mutually exclusive channels; let $x \in [0,1]$ be some relational overlap quantity and $F(x)$ be the closure capacity. Requiring $F(x_1 + x_2) = F(x_1) + F(x_2)$, together with $F(0) = 0, F(1) = 1, F \geq 0$, gives the minimal linear solution $F(x) = x$ under bounded nonnegative conditions, thereby "excluding μ^3 and other nonlinear terms and uniquely locking in the cosine."

This argument does not hold, and the present rewrite corrects it. The flaw lies in an object substitution: Cauchy additivity can indeed force linearity on a nonnegative, decomposable capacity $x \in [0,1]$; but the $\mu = \cos 2(a-b) \in [-1,1]$ on which the correlation depends is not such a capacity that can be decomposed into sums of mutually exclusive nonnegative subparts—it can be positive or negative, and is a relational overlap scalar between two axes. Moving from "F is linear in a nonnegative capacity x" to "K is linear in μ" is an invalid transfer between two different objects.

This can be confirmed directly by a counterexample: there exists a Bell-local model (for example, one with axial triangular-wave correlation $E_{\text{loc}} = -1 + \frac{4}{\pi}d$, where d is the axial angle folded into $[0, \frac{\pi}{2}]$) that satisfies all the structural assumptions in G.6—four-channel nonnegativity, normalization, single-side symmetry (no-signaling), perfect anticorrelation at $a = b$, and capacities still additive over mutually exclusive channels—yet it gives $|S| = 2.0$, not $2\sqrt{2}$. Since models satisfying the same set of structural assumptions can yield both 2 and $2\sqrt{2}$, this set of assumptions (including finite additivity) does not uniquely determine the cosine form. In other words, the "cos" in K_{rs} (rather than a triangular wave or any other $g(\mu)$) is additionally selected; it is directly equivalent to the quantum answer and is not derived from additivity.

Therefore, the honest conclusion is: the axial representation in G.5 derives that "correlation can enter only through $\mu = \cos 2(a-b)$" (legitimate, second class), but the step that "correlation is linearly proportional to μ, namely $E = -\mu$," is a result-level assumption (third class), not a derivation. This corresponds exactly to the situation warned about in the seventh failure criterion of G.17—that the actual dynamics may generate $E = -g(\mu)$ with $g(\mu) \neq \mu$. This appendix no longer pretends that this gap has been closed, but leaves it openly as an unresolved problem.

G.9. From Capacity to Correlation and CHSH (Correct but Trivial Algebra)

Given the cosine form in G.7, the subsequent derivation is elementary and uncontroversial. After freezing, the joint-result correlation value is $AB = rs$, and the readout-layer correlation function is

$$E(a,b) = \sum_{r,s} rsK_{rs}(a,b) = \sum_{r,s} rs \cdot \tfrac{1}{4}(1 - rs\mu).$$

The first term $\frac{1}{4}\sum_{r,s} rs = 0$; the second term is $-\frac{\mu}{4}\sum_{r,s}(rs)^2 = -\frac{\mu}{4} \cdot 4 = -\mu$. Therefore,

$$E(a,b) = -\mu = -\cos 2(a-b).$$

Using the standard polarization CHSH settings $a_0 = 0, a_1 = \frac{\pi}{4}, b_0 = \frac{\pi}{8}, b_1 = -\frac{\pi}{8}$, the four correlation values are respectively $-\frac{1}{\sqrt{2}}, -\frac{1}{\sqrt{2}}, -\frac{1}{\sqrt{2}}, +\frac{1}{\sqrt{2}}$, and the combination is

$$S = E(a_0,b_0) + E(a_0,b_1) + E(a_1,b_0) - E(a_1,b_1) = -2\sqrt{2}$$

$$|S| = 2\sqrt{2}.$$

Emphasis: the correctness of this step does not constitute support for GOCD, because it is entirely inherited from the result-level assumption in G.7. "Assume quantum capacity → obtain quantum CHSH" is tautological. The explanatory step is in G.8, and it remains unfinished.

G.10. No-Signaling (This Step Is a Legitimate Derivation)

Although the joint capacity depends on both gates, the marginal capacity at either single side does not depend on the distant setting. For Alice:

$$\sum_{s} K_{rs}(a,b) = \sum_{s} \tfrac{1}{4}(1 - rs\mu) = \tfrac{1}{2}$$

(since $\sum_s 1 = 2, \sum_s s = 0$), so $P(A = \pm 1 \mid a,b) = \frac{1}{2}$, independent of b. Similarly, Bob's side is independent of a. Therefore $P(A \mid a,b) = P(A \mid a)$ and $P(B \mid a,b) = P(B \mid b)$, that is, operational no-signaling holds.

This step is legitimately derived from the single-side symmetry assumption (G.6), and it is independent of the gap in G.8—regardless of whether the dependence of capacity on μ is cosine or another form, no-signaling follows as long as single-side symmetry holds. Thus, this appendix can honestly claim that global nonseparability does not equal controllable superluminal signaling: the joint correlation depends on the whole closure relation, but neither side can control the observable marginal of the distant side by changing its local setting. This is one of the genuine contributions of this appendix that can stand.

G.11. Why This Is Not Ontic Probability

Formally, $K_{rs}(a,b)$ resembles a joint probability, but in the GOCD context it is not interpreted as "the probability that a single event randomly selects the result (r,s)." It is a closure-channel capacity: under given joint gates, the relational closure capacity that can freeze into (r,s). Single generation remains deterministic—for each actual generative history h, the freezing result $(A,B) = \mathscr{F}(h;a,b)$ is determinate. In repeated readout, the readout layer cannot recover the full generative history and can only record freezing frequencies; thus closure capacity appears as a statistical regularity under a fixed readout protocol. This distinction (deterministic closure at the generation layer versus frequency record at the readout layer) is consistent with Chapter 18, where "probability arises from readout truncation." If future work finds that K_{rs} can only be directly treated as an input probability and cannot be explained through finite generative histories and closure capacity, then this construction fails at this point.

G.12. Finite Discrete Implementation and the Required "Finite Generation Theorem"

To avoid introducing continuous intervals or infinite limits into the generation layer, closure capacity can be implemented as channel counts in finite generative histories: within a fixed readout window, generate L shared generative attempts, each producing a history h_n and a joint freezing $(A_n, B_n) = \mathscr{F}(h_n; a, b)$, and define the channel count

$$N_{rs}(L) = \#\{n \leq L : (A_n, B_n) = (r, s)\}.$$

If the construction holds, there should exist a finite-window error bound

$$\left| N_{rs}(L)/L - K_{rs}(a, b) \right| \leq \varepsilon_L,$$

where ε_L is determined by the readout window, phase fidelity, bandwidth, and closure threshold, rather than deferred to an infinite limit. GOCD should not take $L \to \infty$ as a necessary condition for generation to hold; the more appropriate requirement is that the error be computable, controllable, and capable of failure within a finite window. Therefore, what this appendix truly needs future work to complete is a finite generation theorem: discrete generative histories $\Rightarrow$ four-channel freezing frequencies $\Rightarrow K_{rs}(a, b)$. Until this theorem is completed, K_{rs} remains a conditional closure capacity, not a result derived from the foundational dynamics.

G.13. Equivalent Readout-Gate Algebra Representation (Only a Readout-Layer Shorthand)

For concise expression of the four-channel capacity, one may introduce an equivalent binary gate algebra at the readout layer. It must be emphasized that this algebra is only a finite relational representation of freezing readout channels. It is not a Hilbert space at the generation layer, and it does not mean that GOCD accepts quantum states or projection operators as ontological primitives. Define the axial gate operator

$$Q(a) = \begin{pmatrix} \cos 2a & \sin 2a \\ \sin 2a & -\cos 2a \end{pmatrix}, \quad Q(a)^2 = I, \quad \operatorname{Tr} Q(a) = 0$$

$$Q(a+\pi) = Q(a),$$

and the binary readout gates $G_r(a) = \frac{I+rQ(a)}{2}$, satisfying $G_+(a) + G_-(a) = I$, $G_+G_- = 0$, and $G_r^2 = G_r$. The overlap of two axes is $\frac{1}{2}\operatorname{Tr}[Q(a)Q(b)] = \cos 2(a-b)$. For an anticorrelated source, define

$$K_{rs}(a,b) = \tfrac{1}{2}\operatorname{Tr}[G_r(a)G_{-s}(b)],$$

which expands to $K_{rs} = \frac{1}{4}[1 - rs\cos 2(a-b)]$. This equivalent representation allows the four-channel capacity, completeness, mutual exclusivity, and no-signaling to be expressed uniformly, but it remains only a readout-layer representation. If this matrix form is projected back into the generation layer as a state space, it crosses the ontological boundary of GOCD. Note also that the gate algebra likewise only rewrites the cosine capacity rather than deriving it—it is the same result-level assumption as G.7 in another notation.

G.14. Completed and Unfinished Links (Honest List)

Completed (legitimate derivations): axial gating $\Rightarrow \mu = \cos 2(a-b)$ (G.5); single-side symmetry $\Rightarrow$ no-signaling (G.10); cosine capacity $\Rightarrow E = -\cos 2(a-b) \Rightarrow |S| = 2\sqrt{2}$ (G.9, but only under a result-level assumption).

Unfinished (still assumptions or open problems): first and most crucially, deriving the cosine form $K_{rs} = \frac{1}{4}(1 - rs\cos 2(a-b))$ from GOCD's foundational discrete dynamics; G.8 has shown that it cannot be derived from weak assumptions such as finite additivity, so at present it is a result-level assumption. Second, the dynamical rules by which the measurement gates $G_A(a), G_B(b)$ act on Λ_{AB}. Third, the generative definition of the four joint freezing channels $\mathscr{C}_{rs}$ in terms of events, phase, direction, weight, and closure conditions. Fourth, the computation of K_{rs} from actual finite generative histories, rather than specifying it. Fifth, the generative source by which the axial double-angle relation $(\cos 2a, \sin 2a)$ enters closure. Sixth, the fallible error bound for frequency convergence under a finite read-out window. Until these links are completed, this appendix should be called a "candidate joint-closure construction," not a "complete dynamical derivation of Bell correlations by GOCD."

G.15. Success Criteria

For this candidate construction to be upgraded into substantive mathematical support, it must satisfy the following: first, provide a finite-step update rule $\Lambda_{\lambda+1} = \mathscr{U}_{\text{GOCD}}(\Lambda_\lambda; a, b)$ that produces closure or failure within finite steps; second, generate four mutually exclusive and complete joint freezing channels from that rule; third, derive rather than assume $K_{rs} = \frac{1}{4}[1 - rs\cos 2(a-b)]$—this is especially emphasized compared with the previous version because G.8 has shown that this is precisely where the gap lies; fourth, obtain $E = -\cos 2(a-b)$ and $|S| = 2\sqrt{2}$; fifth, prove no-signaling; sixth, throughout the construction, avoid relying on ontic random sampling, a pre-stored result sample space, setting-dependent postselection, different event ensembles for different settings, infinite limits replacing finite closure, controllable superluminal signaling, or merely renaming quantum probability as closure capacity.

G.16. Failure Criteria

If any of the following occurs, this candidate construction should be regarded as failed or in need of revision: first, the four-channel capacity can only be directly specified and cannot be derived from generative rules (according to G.8, this is precisely the current status—so this item is currently "not yet achieved," rather than "already failed," the distinction being that the dynamical derivation has not yet been fully attempted); second, the so-called closure capacity is in fact only a renaming of quantum probability; third, $|S| > 2$ can be obtained only through setting-dependent filtering, detection postselection, or weight reweighting; fourth, single-side marginals depend on the distant setting; fifth, the model must introduce a pre-stored sample space, a set of possible worlds, or ontic randomness; sixth, closure can be achieved only in infinite steps, continuous limits, or unreachable precision; seventh, the actual discrete dynamics stably produces a correlation of the form $E = -g(\cos 2(a-b))$ with $g(\mu) \neq \mu$ (that is, the cosine is replaced by another monotonic function)—this item directly echoes G.8 and is the most realistic current failure risk of the construction; eighth, the construction cannot distinguish generative failure, readout failure, and low-frequency statistical events. If these failures occur, the book should not protect this construction through narrative restatement, but should acknowledge that GOCD's readout-layer explanation of entanglement remains unfinished on the Bell problem, and may even require revision.

G.17. Relation to the Phase-Fidelity Degradation Law

This appendix handles the stronger task: reproducing CHSH violation under coherence-preserving conditions. It differs from the weaker task: predicting that, when the shared phase reference is damaged, Bell correlations degrade with phase fidelity F_ϕ, $S = S(F_\phi)$, returning near the local bound in the strong-damage limit. The key distinction is that the degradation law does not depend on the gap in G.8, and is therefore the only currently fully falsifiable part of this appendix that does not await the unfinished derivation. But it must be honestly noted that standard decoherence models also predict that correlations weaken as coherence decreases. Therefore, if GOCD is to treat this degradation law as a distinctive prediction, it must provide a quantitative curve distinguishable from standard decoherence, detector noise, or general open-system models, rather than merely stating the qualitative behavior of "monotonic decline." Thus the relation between the two is: the degradation law tests whether readout truncation affects statistical appearance (which can be tested now); the joint-closure construction tests whether GOCD can reproduce CHSH violation itself (awaiting a dynamical derivation). The former cannot replace the latter; only after the latter is completed will the former obtain a stronger model foundation.

G.18. Summary

This appendix provides a GOCD candidate joint-closure construction for Bell/CHSH correlations. It does not attempt to bypass Bell's theorem, nor does it claim that a Bell-local hidden-variable model can violate CHSH—the explicit position is precisely that Bell-local factorization is not satisfied. Its core structure is $(A,B) = \mathscr{C}(\Lambda; a,b)$: the two local readouts are not readings of two pre-stored local properties, but jointly frozen facts after the whole closure of the same shared generative process. In polarization readout, the gate axis $a \equiv a+\pi$ gives the minimal relational representation $\mathbf{q}(a) = (\cos 2a, \sin 2a)$, and therefore $\mu(a,b) = \cos 2(a-b)$ (this step is a legitimate derivation). If the capacity form $K_{rs} = \frac{1}{4}[1 - rs\cos 2(a-b)]$ is further adopted, then $E = -\cos 2(a-b)$ and $|S| = 2\sqrt{2}$ follow, and single-side symmetry guarantees operational no-signaling (the no-signaling step is also legitimately derived).

But this appendix explicitly preserves and emphasizes one unfinished task, now positioned more honestly than before: the cosine form of the capacity is a result-level assumption, not a conclusion derived from structural assumptions. G.8 has shown, using a local counterexample satisfying the same set of assumptions but yielding only $|S| = 2$, that "finite additivity" cannot uniquely force the cosine. Therefore, deriving this capacity from GOCD's foundational phase–weight–direction dynamics is listed as the next necessary constructive task.

The final conclusion of this appendix is therefore twofold, and neither part can be omitted. First, affirmatively: the GOCD general framework, together with a set of explicit, fallible, and institutionally consistent specialized assumptions, is compatible with organizing quantum-type CHSH correlations into a globally nonseparable, operationally no-signaling, no-postselection, no-ontic-probability joint-closure form. This is a real and nontrivial structural statement, and it is consistent with the readout-layer framework of the whole book. Second, negatively and equally importantly: the framework does not automatically derive Bell correlations; the cosine form of quantum correlations is still assumed here rather than derived. Holding both conclusions honestly at the same time, without exaggerating either, is how the book's falsifiable posture is expressed on the hardest problem.

H. Prediction Cards: An Operational Checklist of Falsifiable Outputs

This appendix implements the claim in Part IV (especially Chapter 18) that "probability is readout truncation" as a set of operational prediction cards. Each card follows a unified format: experimental category, closure conditions (the operational preconditions required for the claim to hold), mechanism (generation-layer source), output quantity (measurable proxy and expected behavior), and falsification path (under what controlled change the claim fails). The common discipline of these cards is this: every claim must specify under what controlled parameter change it would be overturned—after probability is downgraded to the readout layer, it obtains, for the first time, the capacity to bear failure (Sections 18.3 and 18.4). All falsification criteria in this appendix belong to the same line of argument as the fourth item of "How to Falsify This Theory" in Part Zero, the Accessibility Proposition in Section 0.3.4, and readout drift in Section 14.5.

H.1. General Card: Probability as Readout Truncation

Experimental Category. Any platform in which readout-protocol parameters (phase fidelity, gating window, bandwidth, threshold) can be systematically changed while the generative state remains unchanged.

Closure Conditions. There exists a determinate, already settled generative state (Section 8.3), on which a controllably adjustable readout protocol is imposed. The three parameters of the readout protocol (bandwidth, gating, and phase fidelity) can be independently changed and independently evaluated.

Mechanism. Probability is not a property of the generative state, but the truncation residue of finite readout applied to determinate generative results (Section 18.1): bandwidth loses phase, gating discretizes time, and threshold discretizes response. The triple structured truncation makes determinate phase structure appear as a statistical distribution, and $|\psi|^2$ appears as the surviving invariant profile under structured truncation.

Output Quantity. Systematic and calculable drift of the statistical distribution with readout-protocol parameters: improving phase fidelity should contract the distribution or reveal previously hidden phase structure; shortening the gating time should reveal phase structure in the statistics; changing bandwidth should produce systematic drift in the distribution.

Falsification Path. If, while the generative state remains unchanged, the readout protocol is systematically changed but the statistical distribution remains strictly unchanged (and remains incompatible after standard decoherence explanations are excluded), then the core claim that "probability arises from readout truncation" fails. This failure is genuine and accepted in advance by this book.

H.2. Double-Slit / Multipath Interference: Visibility Modulation

Experimental Category. Double-slit or multipath interference platforms in which coherence time, detection window, and phase stability can be systematically adjusted.

Closure Conditions. The interference structure corresponds to a determinate generative state; phase fidelity can be independently increased or decreased.

Mechanism. Interference fringes directly express the degree to which phase information is retained in readout; the higher the phase fidelity, the more complete the retained phase structure and the higher the fringe visibility (Sections 18.1 and 18.2).

Output Quantity. Fringe visibility V should increase monotonically with phase fidelity: when phase fidelity is independently confirmed to improve, V must increase monotonically.

Falsification Path. If fringe visibility remains unchanged even when phase fidelity is independently confirmed to have improved, then the explanation that "probability arises from phase truncation" fails on this platform.

H.3. Controlled Decoherence: Statistical Degradation with Coherence Time

Experimental Category. Qubit platforms with adjustable coherence time or controlled decoherence channels.

Closure Conditions. The system is in a determinate generative state; decoherence strength (equivalent to a reduction in phase fidelity and shortening of coherence time τ_φ) can be controlled.

Mechanism. Decoherence is a process in which phase information is structurally truncated by the environment before readout. The shorter the coherence time, the less phase structure is retained, and the more the statistical distribution tends toward phase-free intensity superposition (Sections 6.3 and 18.2).

Output Quantity. Statistical distributions should degrade monotonically with coherence time τ_φ: shortening τ_φ should continuously degrade interference-type statistics into intensity-type statistics.

Falsification Path. If the statistical distribution does not degrade accordingly after coherence time is systematically shortened, then "statistics arises from phase truncation" fails on this platform.

H.4. Uncertainty Relations: As Readout Closure of Fourier Duality

Experimental Category. Wave-packet preparation and detection platforms in which spectral bandwidth and spatiotemporal resolution can be controllably cut.

Closure Conditions. The wave packet corresponds to a determinate generative state; spectral bandwidth can be artificially limited or extended.

Mechanism. In this book, the uncertainty relation is not intrinsic randomness of the world, but the closure relation of "spectrum–spacetime" Fourier duality under the readout protocol (Sections 18.2 and 18.5): bandwidth cutting necessarily causes spreading in the conjugate quantity.

Output Quantity. Systematic spreading of the conjugate distribution with spectral cutting: artificially limiting spectral bandwidth should cause spatial or temporal distributions to spread according to the Fourier-dual relation.

Falsification Path. If the statistical distribution is insensitive to spectral cutting, then the explanation that "uncertainty is Fourier-dual readout closure" fails.

H.5. Bell Correlations: Correlations from a Shared Phase Reference; Reproduction of CHSH Violation Is an Unfinished Task

Compared with the other cards, this card contains one honest downgrading: the book provides an ontological explanation of correlations and a falsifiable degradation law, but does not claim to have derived CHSH violation; the latter is listed as a constructive task with explicit criteria (Appendix G). The items are as follows.

Experimental Category. Controllable phase-coherent platforms in which the shared phase reference can both be maintained and actively disrupted (typical examples include CHSH measurements of entangled photons or spin pairs, qubits with adjustable coherence time, and controlled decoherence channels).

Closure Conditions. The entangled pair shares the same phase reference (operational equivalent: a common coherent resource or common phase framework exists, allowing correlation terms to remain comparable under local gated readout at the two ends). Measurement readout is local gated sampling; no controllable superluminal signaling channel exists; and the phase fidelity of the shared reference can be independently evaluated or systematically lowered through controlled decoherence.

Mechanism. Correlation comes from the projection of a shared phase structure under different measurement bases. The readout ends sample only local discrete results, while statistical correlation is locked by the shared phase reference. Here an unresolved theoretical issue must be honestly marked (Section 18.4): this "shared phase reference" picture is structurally similar to local hidden variables, and Bell's theorem proves that local hidden variables cannot reproduce quantum correlations (CHSH ≤ 2). Whether GOCD falls under the assumptions of Bell's theorem is itself unresolved—because GOCD does not presuppose space or propagation at the generation layer

(Sections 8.2 and 17.4), its correlations may not be spatially local in the first place; however, this book has not yet constructed a discrete phase model that can be proven to produce CHSH > 2. Therefore, the mechanism line of this card claims only the ontological source of correlation (shared phase reference + readout truncation), not that CHSH violation has already been reproduced.

Output Quantity. This card gives two outputs of different strengths, which must be marked separately.

(a) Falsifiable degradation law (the part claimed by this book). When the shared phase reference is actively decohered (by disrupting the shared phase, shortening the effective coherence time, or reducing phase fidelity), the observed Bell-type correlation should continuously degrade as phase fidelity declines, returning near the local bound in the strong-disruption limit. This monotonic degradation relation is an operational and falsifiable prediction (measurable proxy: the monotonic curve of CHSH value versus coherence time/visibility).

(b) Reproduction of CHSH violation (an unfinished task, not a claim of this book). Whether the coherent-retention limit can reach CHSH > 2 (up to the quantum bound $2\sqrt{2}$) is not claimed here as proven. It is a constructive task: one must construct a minimal discrete phase toy model (two bodies, each carrying intrinsic phase and a directional-extension tendency, evolving by difference-update rules and with correlations computed after readout truncation), and numerically test whether its CHSH value can cross 2. This construction is listed in Appendix G. Until it is completed, this card does not cite "CHSH > 2" as an already realized output.

Falsification Path. There are two layers. First, for the degradation law (a): if Bell correlations stably retain their amplitude and remain independent of phase fidelity after the shared phase reference has been explicitly disrupted (or strong phase diffusion has been introduced), after excluding the presence of an unidentified shared reference and common experimental loopholes, then the explanation that "the shared phase reference is the source of correlation" is insufficient and must be replaced by a mechanism that yields a measurable monotonic degradation law. Second, for the constructive task (b): if, after excluding any still-unidentified spatial-locality assumption, no discrete phase toy construction can make CHSH cross the local bound 2, then GOCD's readout-layer explanation of correlations is substantively weakened and should be revised or abandoned accordingly. This capacity to fail is part of the task, not something avoided.

H.6. Measurable Proxies and Order-of-Magnitude Thresholds

The falsification criteria of the cards above are uniformly operationalized through the following measurable proxies: phase fidelity (visibility V, coherence time τ_φ), gating-window width Δt, spectral bandwidth, and the CHSH parameter S. The expectation in each card is expressed as a monotonic or directional response of these proxies under controlled changes, rather than as a match to a particular absolute value, thereby placing the claim under differential testing (consistent with "differences first" in Section 20.5). It must be specially noted that, when CHSH is used as an operational quantity, its use in this appendix must follow the two-layer distinction in H.4: what is falsifiable is the "monotonic degradation law of CHSH with phase fidelity" (output a), whereas "CHSH crossing the local bound in the coherent limit" is an unfinished constructive task (output b, Appendix G), not an already realized operational output.

H.7. General Note: Cards as the Operationalization of Falsifiability

All cards in this appendix share the same logic: compressing the ontological claim that "probability arises from readout truncation" into distributional-difference relations under controlled parameter changes, thereby placing it under direct experimental pressure and making it explicitly negatable. This is precisely the fundamental motivation for downgrading probability to the readout layer in this book—only probability that can fail is physical probability (Section 18.3). It must be specially noted that the Bell card (H.4), compared with the other cards, contains one honest downgrading: it provides an ontological explanation of correlations and a falsifiable degradation law, but does not claim to have reproduced CHSH violation; the latter is listed as the constructive task of Appendix G. This treatment is consistent with Section 18.4 and is an expression, at the level of prediction cards, of the book's discipline: it is better to mark unfinished work than to conceal gaps with narrative.

I. References

These references are divided into eight thematic categories, covering the classical physical frameworks on which this book relies and with which it is in dialogue, simultaneity and causal structure, the intellectual genealogy of discrete spacetime, quantum correlations and decoherence, foundations of information theory, primary philosophical texts, observational data, and computational resources. Where entries are abbreviated, such as multi-paper series, representative works or collections are indicated.

I.1. Preliminary Literature: The Philosophical and Physical Foundations of This Book

The work of this book is built upon the author's nine specific published or in-progress EQT (Energy Quantum Theory) monographs. These works are listed as the concrete physical basis of this book.

- Kaisheng Li and Longji Li. *EQT Foundational Volume: Frequency, Mass, Range of Action, and the Generative Mechanism of Force*. Vienna, Virginia, USA: Gradient Physics Press, 2026. ISBN: 979-8-9961394-1-5.

- Kaisheng Li and Longji Li. *Energy Quantum Theory: A Generative Unified Framework for the Weak Force, the Strong Force, and the Strong–Weak Transition Region*. Vienna, Virginia, USA: Gradient Physics Press, 2026. ISBN: 979-8-9961394-0-8.

- Kaisheng Li and Longji Li. *The EQT Mechanism of Gravitation: From Energy-Quantum Density Gradients to Newtonian Gravity and Weak-Field Geometry*. Vienna, Virginia, USA: Gradient Physics Press, 2026. ISBN: 979-8-9961394-2-2.

- Kaisheng Li and Longji Li. *The EQT Mechanism of Dark Matter and Dark Energy: From Frequency–Density–Gradient to the Dark Components of the Universe*. Vienna, Virginia, USA: Gradient Physics Press, 2026. ISBN: 979-8-9961394-3-9.

- Kaisheng Li and Longji Li. *The EQT Mechanism of Electromagnetic Force: From the Energy-Quantum Density Field to Charge, the Maxwell/QED Limit, and Falsifiability*. Vienna, Virginia, USA: Gradient Physics Press, 2026. ISBN: 979-8-9961394-4-6.

- Kaisheng Li and Longji Li. *EQT Detection Theory: From Observational Ontology to Frequency Matching, Three-State Response, and Falsifiable Predictions*. Vienna, Virginia, USA: Gradient Physics Press, 2026. ISBN: 979-8-9961394-5-3.

- Kaisheng Li and Longji Li. *Frozen Cosmology: The Natural Philosophy of Cosmic Evolution from the Perspective of Energy Quantum Theory*. Vienna, Virginia, USA: Gradient Physics Press, 2026. ISBN: 979-8-9961394-6-0.

- Kaisheng Li and Longji Li. *Mind-Matter Unity: Consciousness and Subjectivity in EQT Process Philosophy of Nature*. Vienna, Virginia, USA: Gradient Physics Press, 2026. ISBN: 979-8-9961394-7-7.

- Kaisheng Li and Longji Li. *Constrained Speculation: The Natural Philosophy of Energy Quantum Theory*. Vienna, Virginia, USA: Gradient Physics Press, 2026. ISBN: 979-8-9961394-8-4.

- Kaisheng Li and Longji Li. *The Generative Structure of Baryons: An Event-Ontological Mechanism for Protons and*

Neutrons in Energy Quantum Theory. Published by Author, 2026. https://doi.org/10.5281/zenodo.18122721

I.2. Primary Physical Literature and Classical Frameworks

- Einstein, A. (1905). Zur Elektrodynamik bewegter Körper. Annalen der Physik, 17(10), 891–921.
- Minkowski, H. (1908). Raum und Zeit. Jahresbericht der Deutschen Mathematiker-Vereinigung (published in 1909; originally a 1908 lecture).
- Einstein, A. (1916). Die Grundlage der allgemeinen Relativitätstheorie. Annalen der Physik, 49, 769–822.
- Planck, M. (1901). Ueber das Gesetz der Energieverteilung im Normalspectrum. Annalen der Physik, 309(3), 553–563.
- Born, M. (1926). Zur Quantenmechanik der Stoßvorgänge. Zeitschrift für Physik, 37, 863–867.
- Heisenberg, W. (1927). Über den anschaulichen Inhalt der quantentheoretischen Kinematik und Mechanik. Zeitschrift für Physik, 43, 172–198.
- Bohr, N. (1928). The quantum postulate and the recent development of atomic theory. Nature, 121, 580–590.
- Robertson, H. P. (1929). The uncertainty principle. Physical Review, 34, 163–164.
- von Neumann, J. (1932). Mathematische Grundlagen der Quantenmechanik. Springer.
- Bohm, D. (1951). Quantum Theory. Prentice-Hall.
- Kronheimer, E. H., & Penrose, R. (1967). On the structure of causal spaces. Proceedings of the Cambridge Philosophical Society, 63(2), 481–501.

- Hawking, S. W., & Ellis, G. F. R. (1973). The Large Scale Structure of Space-Time. Cambridge University Press.

I.3. Simultaneity, Causal Structure, and Definability

- Reichenbach, H. (1928). Philosophie der Raum-Zeit-Lehre. Walter de Gruyter.
- Grünbaum, A. (1963). Philosophical Problems of Space and Time. D. Reidel.
- Winnie, J. A. (1970). Special Relativity without One-Way Velocity Assumptions: Part I. Philosophy of Science, 37(1), 81–99.
- Malament, D. B. (1977). Causal Theories of Time and the Conventionality of Simultaneity. The Journal of Philosophy, 74(8), 481–490.
- Anderson, R., Vetharaniam, I., & Stedman, G. E. (1998). Conventionality of Synchronisation, Gauge Dependence and Test Theories of Relativity. Physics Reports, 295(3–4), 93–180.
- (See also Einstein, A. (1905), in the first category, the original source for the definition of simultaneity in special relativity.)

I.4. Causal Sets and the Intellectual Genealogy of Discrete Spacetime

- Bombelli, L., Lee, J., Meyer, D., & Sorkin, R. D. (1987). Space-time as a causal set. Physical Review Letters, 59(5), 521–524.
- Sorkin, R. D. (1990). Space-time and causal sets. In J. C. D'Olivo et al. (Eds.), Relativity and Gravitation: Classical and Quantum (pp. 150–173). World Scientific.

- Rideout, D. P., & Sorkin, R. D. (2000). A Classical Sequential Growth Dynamics for Causal Sets. Physical Review D, 61, 024002.
- Dowker, F. (2005). Causal Sets and the Deep Structure of Spacetime. In A. Ashtekar (Ed.), 100 Years of Relativity: Space-Time Structure (pp. 445–464). World Scientific.

I.5. Quantum Correlations, Bell Inequalities, and Decoherence

- Bell, J. S. (1964). On the Einstein Podolsky Rosen paradox. Physics, 1, 195–200.
- Clauser, J. F., Horne, M. A., Shimony, A., & Holt, R. A. (1969). Proposed experiment to test local hidden-variable theories. Physical Review Letters, 23, 880–884.
- Clauser, J. F., & Horne, M. A. (1974). Experimental consequences of objective local theories. Physical Review D, 10, 526–535.
- Aspect, A., Grangier, P., & Roger, G. (1982). Experimental Realization of Einstein-Podolsky-Rosen-Bohm Gedankenexperiment: A New Violation of Bell's Inequalities. Physical Review Letters, 49(2), 91–94.
- Hensen, B., et al. (2015). Loophole-free Bell inequality violation using electron spins separated by 1.3 kilometres. Nature, 526, 682–686.
- Zurek, W. H. (2003). Decoherence, einselection, and the quantum origins of the classical. Reviews of Modern Physics, 75, 715–775.
- Joos, E., Zeh, H. D., Kiefer, C., Giulini, D., Kupsch, J., & Stamatescu, I.-O. (2003). Decoherence and the Appearance of a Classical World in Quantum Theory (2nd ed.). Springer.

- Schlosshauer, M. (2005). Decoherence, the measurement problem, and interpretations of quantum mechanics. Reviews of Modern Physics, 76, 1267–1305.

I.6. Information Theory, Bandwidth, and Sampling

- Nyquist, H. (1928). Certain Topics in Telegraph Transmission Theory. Transactions of the AIEE, 47(2), 617–644.
- Kotelnikov, V. A. (1933). On the transmission capacity of the "ether" and of cables in electrical communications (commonly cited through later English translations/collections).
- Shannon, C. E. (1948). A Mathematical Theory of Communication. Bell System Technical Journal, 27, 379–423 & 623–656.
- Whittaker, E. T. (1915). On the functions which are represented by the expansions of the interpolation-theory. Proceedings of the Royal Society of Edinburgh, 35, 181–194.

I.7. Primary Philosophical Texts

- Leibniz, G. W., & Clarke, S. (1717). The Leibniz-Clarke Correspondence. H. G. Alexander (Ed.), Manchester University Press, 1956.
- Kant, I. (1788). Critique of Practical Reason (L. W. Beck, Trans.). Bobbs-Merrill, 1956.
- Mach, E. (1883). Die Mechanik in ihrer Entwicklung: Historisch-kritisch dargestellt. F. A. Brockhaus.
- Whitehead, A. N. (1929). Process and Reality: An Essay in Cosmology. Cambridge University Press.
- Earman, J. (1989). World Enough and Space-Time: Absolute vs. Relational Theories of Space and Time. MIT Press.

- Dieks, D. (Ed.). (2006). The Ontology of Spacetime. Elsevier.
- Ladyman, J., & Ross, D. (2007). Every Thing Must Go: Metaphysics Naturalized. Oxford University Press.

I.8. Observational Data and Experimental Compilations

- LIGO Scientific Collaboration and Virgo Collaboration. (2016). Observation of Gravitational Waves from a Binary Black Hole Merger. Physical Review Letters, 116, 061102.
- Planck Collaboration. (2020). Planck 2018 results. VI. Cosmological parameters. Astronomy & Astrophysics, 641, A6.
- Particle Data Group. (2022). Review of Particle Physics. Progress of Theoretical and Experimental Physics, 2022(8), 083C01.
- Santini, P., et al. (2023). CEERS-related series of papers (ApJL 949 series, multiple papers).

I.9. Open-Source Computational Resources

- Kluyver, T., et al. (2016). Jupyter Notebooks—A publishing format for reproducible computational workflows. In F. Loizides & B. Schmidt (Eds.), Positioning and Power in Academic Publishing (pp. 87–90). IOS Press.
- Harris, C. R., et al. (2020). Array programming with NumPy. Nature, 585, 357–362.
- Virtanen, P., et al. (2020). SciPy 1.0: Fundamental Algorithms for Scientific Computing in Python. Nature Methods, 17, 261–272.

www.ingramcontent.com/pod-product-compliance
Lightning Source LLC
LaVergne TN
LVHW052021200926
848481LV00005B/1549